Scenario Planning and Tourism Futures

THE FUTURE OF TOURISM
Series Editors: Ian Yeoman, *NHL Stenden University of Applied Sciences, the Netherlands* and **Una McMahon-Beattie**, *Ulster University, Northern Ireland, UK*

Some would say that the only certainties are birth and death; everything else that happens in between is uncertain. Uncertainty stems from risk, a lack of understanding or a lack of familiarity. Whether it is political instability, autonomous transport, hypersonic travel or peak oil, the future of tourism is full of uncertainty but it can be explained or imagined through trend analysis, economic forecasting or scenario planning.

This new book series, The Future of Tourism, sets out to address the challenges and unexplained futures of tourism, events and hospitality. By addressing the big questions of change, examining new theories and frameworks or critical issues pertaining to research or industry, the series will stretch your understanding and generate dialogue about the future. By adopting a multidisciplinary perspective, be it through science fiction or computer-generated equilibrium modelling of tourism economies, the series will explain and structure the future – to help researchers, managers and students understand how futures could occur. The series welcomes proposals on emerging trends and critical issues across the tourism industry and research. All proposals must emphasise the future and be embedded in research.

All books in this series are externally peer-reviewed.

Full details of all the books in this series and of all our other publications can be found on http://www.channelviewpublications.com, or by writing to Channel View Publications, St Nicholas House, 31-34 High Street, Bristol, BS1 2AW, UK.

THE FUTURE OF TOURISM: 10

Scenario Planning and Tourism Futures

Theory Building, Methodologies and Case Studies

Albert Postma, Stefan Hartman and Ian Yeoman

CHANNEL VIEW PUBLICATIONS
Bristol • Jackson

DOI https://doi.org/10.21832/POSTMA8878
Library of Congress Cataloging in Publication Data
A catalog record for this book is available from the Library of Congress.
Names: Postma, Albert, author. | Hartman, Stefan, author. | Yeoman, Ian, author.
Title: Scenario Planning and Tourism Futures: Theory Building, Methodologies and Case Studies/Albert Postma, Stefan Hartman and Ian Yeoman.
Description: Bristol, UK; Jackson, TN: Channel View Publications, 2025. | Series: The Future of Tourism: 10 | Includes bibliographical references and index. | Summary: 'This book offers a practical guide for scenario planning to make sense of the future of tourism for practitioners, researchers and students. A variety of case studies are presented which demonstrate how scenario planning is deployed. They include learning points and questions to help readers understand key concepts, theories and applications' – Provided by publisher.
Identifiers: LCCN 2024024677 (print) | LCCN 2024024678 (ebook) | ISBN 9781845418861 (paperback) | ISBN 9781845418878 (hardback) | ISBN 9781845418892 (epub) | ISBN 9781845418885 (pdf)
Subjects: LCSH: Tourism – Planning – Case studies. | Tourism – Forecasting – Case studies.
Classification: LCC G155.A1 P679 2025 (print) | LCC G155.A1 (ebook) | DDC 910.68/4 – dc23/eng/20240719
LC record available at https://lccn.loc.gov/2024024677
LC ebook record available at https://lccn.loc.gov/2024024678

British Library Cataloguing in Publication Data
A catalogue entry for this book is available from the British Library.

ISBN-13: 978-1-84541-887-8 (hbk)
ISBN-13: 978-1-84541-886-1 (pbk)

Channel View Publications
UK: St Nicholas House, 31-34 High Street, Bristol, BS1 2AW, UK.
USA: Ingram, Jackson, TN, USA.

Website: https://www.channelviewpublications.com
X: Channel_View
Facebook: https://www.facebook.com/channelviewpublications
Blog: https://www.channelviewpublications.wordpress.com

Copyright © 2025 Albert Postma, Stefan Hartman and Ian Yeoman.

All rights reserved. No part of this work may be reproduced in any form or by any means without permission in writing from the publisher.

The policy of Multilingual Matters/Channel View Publications is to use papers that are natural, renewable and recyclable products, made from wood grown in sustainable forests. In the manufacturing process of our books, and to further support our policy, preference is given to printers that have FSC and PEFC Chain of Custody certification. The FSC and/or PEFC logos will appear on those books where full certification has been granted to the printer concerned.

Typeset by Riverside Publishing Solutions.

Contents

Figures and Tables — ix
About the Authors — xiii
Foreword 1 — xv
Falco de Klerk Wolters

Foreword 2 — xvii
Jeremy Sampson

1 Why the Future is Important in an Uncertain World: The Role of Scenario Planning to Create the Future of Tourism — 1
  1.1 A World in a Constant State of Change — 1
  1.2 Dealing with Uncertainties: Scenario Planning and Strategic Foresight — 2
  1.3 The Birth and Growth of the European Tourism Futures Institute — 3
  1.4 Concluding Remarks: Twelve Years of Best Practice — 5
  1.5 Discussion Questions — 7

2 Is Scenario Planning a Theory Builder or a Research Methodology? — 8
  2.1 Introduction — 8
  2.2 What is Future Studies? — 8
  2.3 What are Foresight Methods? — 9
  2.4 What is Scenario Planning (or Scenarios)? — 10
  2.5 Scenario Planning: Theory or Methodology? — 11
  2.6 Methodological Domination — 13
  2.7 Theory Building — 14
  2.8 Scenario Planning as a Theoretical Framework — 15
  2.9 Scenario Planning as a Learning Framework — 17
  2.10 The ETFI Method as a Framework of Theory and Practice — 18
  2.11 Concluding Remarks — 21
  2.12 Discussion Questions — 21

3 Strengthening Resilience in Tourism through Scenario Planning 22
   3.1 Why Destination Resilience? 22
   3.2 Resilience of What? 23
   3.3 Resilience to What? 23
   3.4 Resilience by Whom? 27
   3.5 Concluding Remarks 28
   3.6 Discussion Questions 29

4 Methodologies of Scenario Planning and Strategic Foresight 30
   4.1 The Urge to Visualise the Future 30
   4.2 The Birth of Scenario Planning (First Era) 31
   4.3 Second Era of Scenario Planning 33
   4.4 Third Era of Scenario Planning – Consolidation of Scenario Planning 34
   4.5 Strategic Foresight 35
   4.6 The Use of Strategic Foresight 36
   4.7 Strategic Foresight Process and its Components 37
   4.8 Scenarios as a Key Tool in Strategic Foresight 38
   4.9 Outcome of Strategic Foresight 38
   4.10 Concluding Remarks 39
   4.11 Discussion Questions 40

5 The Application of Scenario Planning in Tourism 41
   5.1 Introduction 41
   5.2 Case Study One: The Policy and Politics of New Zealand Tourism 42
   5.3 Case Study Two: Preparing for a Crisis 45
   5.4 Case Study Three: Trends Analysis and Product Development: The Importance of Family Tourism During COVID-19 47
   5.5 Case Study Four: Using Scenario Planning and Creativity to Design Future Hotels 50
   5.6 Case Study Five: Using Scenario Planning to Explore the Future of Work Through Technology 53
   5.7 Concluding Remarks 57
   5.8 Discussion Questions 58

6 The European Tourism Futures Institute Method 59
   6.1 The Futures Cone 59
   6.2 Formulating and Delimiting the Strategic Question 59
   6.3 Developing a Predictive Scenario 60
   6.4 From Baseline to Alternative Scenarios 62
   6.5 Developing Explorative Scenarios 62
   6.6 From Scenarios to Strategy 78
   6.7 Concluding Remarks 83
   6.8 Discussion Questions 84

7	A Scenario Framework for the Post COVID-19 Futures of Tourism	85
	7.1 Background	85
	7.2 Approach	86
	7.3 Outcomes of the Study	96
	7.4 Concluding Remarks	96
	7.5 Discussion Questions	98
8	The Tourism Futures of Rural Friesland: An Integrated Spatial Planning Approach to Tourism Planning	99
	8.1 Background	99
	8.2 Approach	100
	8.3 Outcome	120
	8.4 Concluding Remarks	122
	8.5 Discussion Questions	123
9	Futures Lab Fryslân	125
	9.1 Background	125
	9.2 Future Challenges	128
	9.3 Purpose	128
	9.4 Concluding Remarks	144
	9.5 Discussion Questions	145
10	Scenarios for Inbound Tourism to the Netherlands – Case Study	147
	10.1 Purpose	147
	10.2 Background	147
	10.3 Approach	148
	10.4 Concluding Remarks	161
	10.5 Discussion Questions	161
11	Notting Hill Carnival Futures 2020	162
	11.1 Background	162
	11.2 Purpose	164
	11.3 Approach	164
	11.4 Concluding Remarks	180
	11.5 Discussion Questions	184
12	Visitor Pressure in European Cities	186
	12.1 Background	186
	12.2 Purpose	188
	12.3 Approach	188
	12.4 Concluding Remarks	203
	12.5 Discussion Questions	205

13 The European Tourism Futures Institute on the Edge of Time   207
    13.1 Exploration Phase: The Initiative for the European
          Tourism Futures Institute (2007–2011)   207
    13.2 Involvement Phase (2011–2014)   208
    13.3 Development Phase (2015–2018)   208
    13.4 Consolidation Phase (2018 to now)   213
    13.5 What Next?   214
    13.6 Concluding Remarks   218
    13.7 Discussion Questions   219

    References   220

    Index   233

Figures and Tables

Figures

2.1	Learning loop	19
3.1	Out of balance	25
3.2	Re-balance	25
3.3	Tipping point	26
3.4	Bounce back and bounce forward	27
4.1	Components of strategic foresight	37
5.1	Scenario planning matrix – New Zealand	44
5.2	Trends matrix	48
5.3	Cartoon style diagram – LifeStyle Hub	52
5.4	Scenario planning matrix – Future of Work	56
6.1	Futures cone – three types of scenarios compared	60
6.2	Strategic foresight process	63
6.3	Visualisation of the outcome of horizon scanning	65
6.4	Example horizon scan with four representatives from city DMOs in Europe	66
6.5	Visualisation of clustering of the outcomes of the horizon scan into key processes	67
6.6	Visualisation of driving forces identification	68
6.7	Visualisation of driving force's limits of the plausible	69
6.8	Visualisation of impact by uncertainty matrix	70
6.9	Visualisation of the shifting and sorting of the driving forces	71
6.10	Three types of explorative scenarios	72
6.11	Example of a trilemma triangle	73
6.12	Scenarios are framed by the extremes of the two axes in the scenario cross	74
6.13	Scenario 'Natural Wadden Sea Dike' in a picture (made by JAM Visual thinking) (project 2015)	78
6.14	Four scenarios for tourism in the province of Limburg, Netherlands, in pictures	79
6.15	Implication tree (forward thinking) or futures wheel (backward thinking)	80

7.1	Scenario framework for post-COVID tourism	87
8.1	Scenario planning process	101
8.2	Importance by uncertainty matrix	103
8.3	Scenario framework	103
8.4	Scenario 1	105
8.5	Scenario 2	109
8.6	Scenario 3	113
8.7	Scenario 4	117
9.1	Location of the province of Friesland in the Netherlands	126
9.2	Strategic foresight defined	130
9.3	Advertisement of MURAL on internet	133
9.4	Future scanning instructions	134
9.5	The environment of an organisation	135
9.6	Importance by uncertainty matrix	136
9.7	Scenario framework with illustrations	140
9.8	A taste of the four scenarios: Headlines of four newspaper articles	141
9.9	Summary of each scenario in its own font type	142
9.10	Key features of each scenario	143
10.1	Domain map of the study	150
10.2	Era analysis of inbound tourism to the Netherlands	153
10.3	Scenario cross	157
11.1	Impression of crowds at Notting Hill Carnival	163
11.2	The logo of the Notting Hill Carnival Futures 2020 project	164
11.3	Impression of performance at Notting Hill Carnival	165
11.4	Impression of performance at Notting Hill Carnival	166
11.5	Participants from the Notting Hill Carnival community sharing their experiences	167
11.6	Core values with dot votes representing appreciation (green stickers) and uncertainty/doubt (red stickers) concerning the future of the Notting Hill Carnival	168
11.7	Impression of performance at Notting Hill Carnival	169
11.8	Perceived implications if the core values would be nurtured or ignored	170
11.9	Composition of Notting Hill's Carnival Public	170
11.10	Core values with dot votes for importance (green stickers) and worries about the future (red stickers) by visitors of the Notting Hill Carnival	171
11.11	Two USPs identified by the participating visitors	172
11.12	Festival network organisations	173
11.13	Experts of festival network organisations at work	173
11.14	Four scenarios for Notting Hill Carnival Futures 2020	175
11.15	Impression of performance at Notting Hill Carnival	177
11.16	Impression of Notting Hill Carnival	178

11.17	Impression of Notting Hill Carnival	181
11.18	Overall participant satisfaction with workshop aspects	183
11.19	Impression of performance at Notting Hill Carnival	183
12.1	Driving forces of change and critical uncertainties for city tourism in 2025 (round 1)	190
12.2	Four scenarios for the development of city tourism (round 1)	190
12.3	Driving forces of change and critical uncertainties for city tourism in 2025 (round 2)	195
12.4	Example of two clusters of driving forces that were created during the workshop	195
12.5	Four scenarios for the development of city tourism (round 2)	195
12.6	BSR lifestyle model	199
12.7	Opening questions in an online tool to structure the debate on overtourism using scenario planning	204
12.8	Results from an online tool to structure the debate on overtourism using scenario planning	205

Tables

2.1	Types of forecasts	17
3.1	Shocks and stresses	24
4.1	Comparison between predictive and explorative scenarios	32
6.1	Steps in the adaptive strategic foresight process	64
6.2	Possible sources of horizon scanning	66
9.1a	Number of visitors to province of Friesland, by origin per year (*1000)	127
9.1b	Origin of international tourists to province of Friesland, per year (*1000)	127
9.2	Long list of driving forces with the highest perceived level of impact	136
9.3	First idea of key uncertainties	137
10.1	Domain analysis of the study	149
10.2	Overview of the forcefield that impacts upon inbound overnight stay tourism to the Netherlands	152
11.1	Core values of the Notting Hill carnival, as perceived by the participants	168
11.2	Unique Selling Points of the Notting Hill Carnival as perceived by a group of visitors	172
12.1	Robustness of the strategies for the four scenarios	201
12.2	Opportunities and threats of the four scenarios	202
12.3	Evaluation criteria of potential strategies	203
13.1	International keynotes and performances about scenario planning and scenarios in tourism	209
13.2	Scenario planning projects conducted by the ETFI	211

From Albert:
*To our university and industry partners for their
confidence in the quality of our work*

From Stefan:
*The team of the European Tourism Futures Institute,
for their dedication, support and inspiration*

From Ian:
*Hugo Sunderland, the labradoodle that inspired
the future of tourism*

About the Authors

Albert Postma is Professor of Strategic Foresight and Scenario Planning at the European Tourism Futures Institute (ETFI – www.etfi.eu), NHL Stenden University in The Netherlands, where he is responsible for the learning line of this domain across the curriculum in the B Leisure Management and B Tourism Management programme. Postma holds an MSc and PhD in spatial Sciences. In his PhD study he investigated tipping points in the attitude of residents towards tourism in their home communities. This thesis paved the way for overtourism studies in various European cities and a few early academic articles, which put him at the forefront of the overtourism debate. Postma's current research focuses on strategic foresight, scenario planning next to tourism community relations (overtourism). Postma is a respected speaker at business conferences, has authored dozens of technical reports and articles, and is co-editor of the *Journal of Tourism Futures* and the book *The Future of European Tourism* (2013).

Stefan Hartman holds the position of Head of Department of the European Tourism Futures Institute (ETFI – www.etfi.eu) at NHL Stenden University, Leeuwarden, The Netherlands. At the ETFI he helps actors in the leisure and tourism industry to develop strategies and actions that allow them to manage continually changing business environments. To do so, he uses his knowledge of transition management, resilience and adaptive capacity building. Stefan obtained his PhD at the University of Groningen, The Netherlands. His research focuses on the development, strategic (spatial) planning and governance issues related to spaces and places that are in the process of becoming (smart) destinations for tourism and leisure.

Ian Yeoman (LPSNZ) is Professor of Disruption, Innovation and New Phenomena at NHL Stenden University of Applied Sciences, The Netherlands. Ian is the champion of tourism futures based upon his initial work as the Scenario Planner at VisitScotland where he introduced scenario construction, economic modelling, and trends analysis within the organisation to understand and make sense of the external environment. Ian then moved to New Zealand where he was involved in several tourism futures projects for government agencies and national strategies while an Associate Professor at Victoria University of Wellington. Ian is the co-editor of the *Journal of Tourism Futures*, editor of the *Journal of Revenue and Pricing Management* and co-editor of The Future of Tourism book series. Ian has a PhD in Operations Research from Edinburgh Napier University and is the author of over 70 research papers and 22 books. Forthcoming titles include *2075 – The Future of Food Tourism*, *The Future of Hotels* and *Scenarios for Global Tourism*. He is also a passionate Sunderland AFC fan (so don't mention Newcastle United!).

Foreword 1

15 Years of the European Tourism Futures Institute (ETFI)

The official launch of the European Tourism Futures Institute (ETFI) at the Emmen Zoo in 2009 is now almost 15 years ago. The ETFI experienced a successful start-up phase (2009–2013) which laid the foundation in 2023 for the current research group with a well-established reputation in the Netherlands, as well as internationally.

Around 2009, higher education in the Netherlands, and NHL Stenden University in particular, was looking for new initiatives to contribute to the knowledge economy. This happened at the time of an economic recession and at the heights of the international 2008–2009 financial crisis. Initiatives that related to knowledge development and transfer were very much welcomed by governments.

At universities of applied science, these initiatives of knowledge transfer between education, research and industry were situated within educational faculties, sometimes also referred to as the New Faculty (and no longer as separate institutes). As director of the School of (later the Academy of) Leisure and Tourism from 2008 to 2022, I saw the opportunities and made use of this situation.

Why did we decided to focus on scenario planning? The idea originated on the work floor, as a result of talks between colleagues, discussing what would be really innovate for the tourism and leisure industry and education. Especially when Dr Elena Cavagnaro, lector Sustainability, told me that long-term strategic thinking would be an added benefit in the sustainability discourse and when my colleague Dr Kenneth Miller told me about the promising scenario studies from Dr Ian Yeoman at VisitScotland, after foot and mouth disease made tourism collapse in Scotland. So, I travelled to Edinburgh and got inspired by Ian's work and ideas. And the rest is history. Looking back now after 15 years, the choice to focus on scenario planning has proved to be a success.

With the help of the industry associations Hiswa (watersports) and RECRON (accommodation providers), support from ANVR and NRIT and international support from, among others, the University of Michigan, we were able to get funds from the various governmental authorities located in the Northern part of Netherlands (Province of Fryslân, Province of Drenthe, municipality of Leeuwarden, municipality of Emmen, subsidy program 'Samenwerkingsverband Noord-Nederland' (SNN)) and the EU (EFRO program) for an initial period of four years (2009–2013). The aim was, and still is, quite simple: to contribute to the academic discourse in scenario planning and strategic foresight, to contribute to professional practice of tourism and leisure in the northern Netherlands and in other regions in Europe and to contribute to state-of-the-art education in leisure and tourism at bachelor level (International Tourism Management and Leisure and Event Management) and master level (International Leisure, Tourism and Event Management).

The first ETFI book, entitled *The Future of European Tourism*, edited by Dr Albert Postman, Dr Ian Yeoman and Dr Jeroen Oskam, was a manifesto of that first period and its successes. Now, the second ETFI book is here! A handbook for students, educational professionals, research staff and practitioners who want to work with scenario planning and strategic foresight, so as to understand how useful these methods are in shaping a better future. Scenario planning is a fantastic method, since it helps you to understand uncertainties through identifying patterns of change as insights and then addressing what the future might be as foresight.

Falco de Klerk Wolters
Former Director of the Academy of Leisure & Tourism
NHL Stenden University

Foreword 2

Focus on the Future of Tourism

Although I'm often described as working in the field of 'sustainable tourism', I rarely use that term myself. I find it more relevant and interesting to say that my work, and that of Travel Foundation, concerns the future of tourism, which entails supporting industry stakeholders to prepare for the future, address priority risks and opportunities and generate a more balanced visitor economy.

The future of tourism is, of course, uncertain and growing more so by the minute. Megatrends, such as climate change, continued growth in tourist numbers and advances in technology will have an outsized yet unpredictable effect on the sector. However, one thing *is* certain: business as usual is the least likely outcome.

Our recent report with ETFI and partners, 'Envisioning Tourism in 2030 & Beyond', starkly sets out the incompatibility of the current tourism model within a decarbonising global economy – the shape of travel will have to change significantly and there are many signs that this transition is already underway. Destinations will need to adapt to changing climate impacts – with visitor patterns likely to be every bit as uncertain as weather patterns. Some may choose to get on the front foot and lean towards innovation and new opportunities, while others may wait to have change imposed on them. But change is coming – ready or not.

And that's where scenario planning comes in. It is critical that destinations prepare for change and study the implications of different variables in order to focus on the 'best case scenario' outcome for local communities. The most successful organisations of the future will understand the risks and opportunities ahead and the levers that can turn possible scenarios into likely ones. This knowledge will help immensely with decision making around planning, policies, product development and the allocation of resources.

How we prioritise our efforts and identify the most urgent actions is critical to our future, but the tourism sector has a poor track record of doing this well. Therefore, this book is particularly welcome for providing examples where scenario planning has been used to great effect. The authors demonstrate how this is one of the best tools we have for engaging a wide variety of stakeholders in evidence-based decision making and formulating effective action plans for businesses and governments – ideally both in collaboration. Scenario planning is an antidote for tourism's short termism, kneejerk reactions and unintended consequences. Future thinking is the root of sustainability and we must, therefore, ensure the skills, data and tools for this discipline are commonplace at every level within the sector and will remain so for future generations.

Jeremy Sampson
Chief Executive Officer, The Travel Foundation
Founding Chair (2020–2023), Future of Tourism Coalition

1 Why the Future is Important in an Uncertain World: The Role of Scenario Planning to Create the Future of Tourism

Learning Points
- The future for tourism businesses, organisations and destinations is complex, dynamic and uncertain.
- The challenge is to become less fragile and more future proof.
- Strategic foresight with the help of scenario planning is a key approach.
- The European Tourism Futures Institute is the leading institution to future proof the tourism industry with the help of scenario planning.

1.1 A World in a Constant State of Change

In 2019, the COVID-19 pandemic broke out, with huge consequences for travel and tourism around the globe. Lockdowns and other (government) measures have fuelled and surfaced the discontent of large groups of citizens and has, in some parts of society, boosted the decline of the respect for laws and regulations. While the COVID virus continued to proliferate in ever-changing guises, with infections coming and going in waves, the apparent need to go on holiday became so severe that demand for holiday making exploded like a pressure cooker during the summer of 2022. Since many staff left their job in travel and tourism during the heaviest period of the COVID-19 pandemic, due to the lack of work, closed airports and largely absent train traffic, the industry in Europe faced huge difficulties in coping with the crowds.

In 2022, Russia started its war against Ukraine, creating all sorts of tensions for tourism. It led to additional large migration flows, for which accommodation capacity in European countries often proves insufficient. The energy supply of gas within Europe and the world food supply of grain have been endangered by the war. Prices for oil and gas went up in unprecedented ways, creating cashflow problems, particularly for the micro-, small- and medium-sized companies, some of which went bankrupt. With new membership applications in eastern Europe, the European Union will have to deal with a shift in emphasis towards the east, as well as the geopolitical consequences within Europe and the European Union itself as an institution.

Climate change, and the measures that are needed in response, is also among the tensions, both in the short and the long term. Climate change contributes to (too) high temperatures and many nature fires in well-known holiday areas in Spain, Portugal, France, Italy, Greece, etc. In the high mountains, the 'eternal snow' is disappearing, and glaciers are retreating, causing peaks in meltwater run-off in the short run and in the long run in shortages of water for the large rivers that supply large parts of Europe with fresh water. Because of rises in the sea level, many destinations around the globe are at risk of disappearing in a few decades time. The climate in several parts of the world may become so extreme that it makes places less attractive to visit, contributes to a redistribution of tourism flows, as well as confronting some regions with an increase of immigrants searching for better inhabitable places. To fight climate change, the pressure to make tourism more sustainable is increasing. The Glasgow Declaration, which has been signed by over 500 tourism organisations, aims to halve emissions by 2030 and reduce it to zero by 2050. To meet the goals and aims of the Glasgow Declaration, a range of substantial interventions are needed (Peeters & Papp, 2023), such as reducing the number of flights, taxing fossil fuels, investing in alternative emission free fuels and improving the (fast) train network in Europe.

Clearly, we live in a 'VUCA world': volatile, uncertain, complex and ambiguous (Hartman, 2023; Lubowiecki-Vikuk & Sousa, 2021). This means we need to accept the situation of constant, unpredictable change and a lack of clarity about how interconnected factors are shaped and changed and how these are creating an environment that is difficult to foresee, interpret and analyse (Major & Clarke, 2022).

1.2 Dealing with Uncertainties: Scenario Planning and Strategic Foresight

Developments and events beyond our control, such as those mentioned above, have not or have insufficiently been anticipated and responses are pretty much dominated by Actors, often using the excuse 'if we would have the knowledge of today, we would have acted differently'. Consequently, they have difficulties in dealing with them properly, dissatisfaction among citizens is growing, belief in and respect for science, laws and rules is decreasing, people are increasingly taking their own destiny into their own hands and conspiracy thinking, populist and right-wing extremist ideas are on the rise.

Developments, such as discussed above, are not only unexpected, inconvenient or even dissatisfactory, they may also be seen as factors bringing the complex tourism system out of balance and as indications of a systemic crisis (Hartman, 2021; Postma & Yeoman, 2021). In such circumstances, crisis management seems the only way left to try to fight the uncertainties, featured by short-term thinking, re-active behaviour and fighting the fire instead of preventing it. Actors often apologise by saying 'if we would have the knowledge of today, we would have acted differently'. Since tourism and its industry is vulnerable to changes within the system, whether they are disruptive or sustaining, it is preferable to think long term and to act pro-actively instead – to signal changes in the system early, to explore the possible positive and negative consequences of such imminent changes and to take measures before it is too late.

Grounded in the belief that change is always possible, scenario planning and strategic foresight offer an appropriate thinking and working approach that establish knowledge of the future with the possibility to act now! In this view there is not one single inevitable future, but a multiplicity of futures that need to be taken into account. Scenario planning and strategic foresight help to envision extreme yet plausible futures and to anticipate these futures pro-actively before they may occur, by formulating strategies or policies that facilitate destinations, businesses or organisations to become more resilient or 'future proof' to such circumstances.

1.3 The Birth and Growth of the European Tourism Futures Institute

In the late 2000s, when the Food and Mouth disease broke out, and with potentially severe consequences for tourism around the globe, the Academy of Leisure and Tourism at NHL Stenden University, and a number of stakeholders in tourism, took the initiative to establish a European Top Institute that could support the sector by looking ahead and anticipating such events. So far, the emphasis of the Academy's research activities (since 1988) had been predominantly on monitoring and evaluation (with monitoring system of Toerdata, as an example) – looking back. The ambition of the new institute was to look forward: to establish a better understanding of the future, to support the industry in dealing with future uncertainties and to shape the preferred future. With the aim to contribute to the innovation of the sector, and indirectly to the economy and wellbeing of citizens in tourism destinations, by means of (a) collecting and disseminating knowledge about trends and developments that drive the change, (b) inspiring businesses and organisation to consider these drivers with reference to their own future and to ignite creative processes to think outside the box and (c) scientific and applied research to monitor the development of drivers of change and strategic responses, and the effects thereof on the longer term. Given the tourism industry's lack of knowledge about the future, the lack of instruments and tools to study tourism's future and the tourism industry's lack of know how to transfer research outcomes of the future into strategies and policies, the European Commission decided to support the initiative for four years.

Inspired by Ian Yeoman's scenario planning work at VisitScotland, the ambitions resulted in the creation of The European Tourism Futures Institute (ETFI) in 2009. The ETFI started its operations in the autumn of 2010 with the appointment of a professor of scenario planning and a personal assistant, while in early 2011 a programme manager was also appointed. The small team saw the upcoming ITB[1] in Berlin – the largest tourism fair in the world – in March 2011 as an opportunity to establish its name. After just a few weeks, a brand logo and house style were developed. The ETFI managed to act as a key sponsor of the Futures Days at the ITB conference,[2] which gave it a prominent place during the conference with its logo presented on big banners, a leaflet on each of the hundreds of chairs in the conference hall, access to the VIP-lounge and a performance at the main stage.

During the first four years the ETFI managed to establish a growing portfolio of public and private clients. Initially, it was mainly in the surrounding region of North Netherlands, but increasingly in the rest of the country and internationally as well. The ETFI conducted both self-initiated studies and projects commissioned by clients, with Ian Yeoman as supporter and mentor. While the work of the ETFI was increasingly acknowledged by the tourism industry, the ETFI also established an academic journal in the new and unexplored niche of tourism futures: the *Journal of Tourism Futures*.

While the subsidies of ETFIs first four years had come to an end, the ETFI saw an expansion in the number and the scope of projects, a growth in the demand for (keynote) presentations about scenario planning, strategic foresight and specific projects, a rise in the reputation of the *Journal of Tourism Futures* and an endurance of educational relations with other institutions. The ETFI got involved and associated with other key players in tourism such as the Organisation for Economic Cooperation and Development (OECD), the World Tourism Organization (UNWTO), the World Tourism and Travel Council (WTTC), the European Travel Commission (ETC), European Cities Marketing (ECM, now called City Destinations Alliance), European Association of European Tour Operators (ETOA) and the Travel Foundation (TF).

1.4 Concluding Remarks: Twelve Years of Best Practice

This book reflects nearly 15 years of the ETFI's best practice. It displays the ETFI's vision of and approach to scenario planning and strategic foresight. Not only theoretically but also by means of a number of case studies that demonstrate the diversity of industry challenges concerning the future and the ETFI's approach to dealing with them. The book, published in Channel View's The Future of Tourism series, is the first that is dedicated to scenario-planning methodologies and strategic foresight in the context of tourism, hospitality and events. The book is timely, as, given the disruptive changes of technology, climate change and COVID-19 in a tourism context, many destinations, students, academics and practitioners are trying to make sense of this new world. The book brings order in a sense making process using scenario planning and strategic foresight to build resilience and contribute to future proofing. It is intended as a reference book for scenario planning and strategic foresight methodology in tourism, hospitality and events.

The purpose of this book is twofold. It aims to use scenario planning and strategic foresight as a theory builder in tourism futures research for researchers, and it aims to offer a practical guide 'how to do' scenario planning, in order to make sense of the future of tourism. Therefore, the book is especially interesting for postgraduate researchers and students involved in strategic planning, scenario planning and tourism futures. For practitioners it offers a how-to guide to scenario planning in the tourism industry with examples and learning points.

Chapters 2–6 of *Scenario Planning and Tourism Futures*, overview the theoretical development of scenario planning and describes the scenario planning method and relates it to strategic foresight. In Chapter 2, a theory building perspective will be presented. Future studies will be discussed from an ontological and epistemological perspective, core concepts and theories associated with future studies will be highlighted and scenario planning will be presented as a conceptual framework to inform theory building. Chapter 3 discusses the concept of resilience in relation to tourism destinations as complex systems. In Chapter 4, the methodology of scenario planning and strategic foresight will be outlined. The notion of a single and predictable future will be compared and contrasted with the conception of multiple and unpredictable futures. Scenario development will be explained as a tool to map multiple futures and strategic foresight as the methodology to link scenarios with strategy. Chapter 5 discussed the application of scenario planning in tourism. It illustrates various contexts in which scenario planning can be applied. Chapter 6 shows how the methodology of scenario planning and strategic foresight have been developed by the ETFI into an applicable method. Each subsequent step in the process will be explained and illustrated.

Chapters 7–11 contain a variety of case studies in which the deployment and use of the ETFI's approach to scenario planning and strategic foresight is demonstrated. Each case study is drawn from the European Tourism Futures Institute's 15 years of practice, emphasising the different stages of the scenario planning method. Each chapter includes a series of discussion questions for readers to help them understand key concepts, theories and application.

In Chapter 7, a scenario framework for the post-COVID-19 future of tourism will be presented. It is based on a self-initiated scenario study conducted in April/May 2020, shortly after the outbreak of COVID-19 in Europe. The case study presents a framework of four scenarios on the future of tourism in the post-COVID era. The scenarios have remained valid for quite some time already and have been applied in various contexts (destinations, organisations and for planning and policymaking). Moreover, the scenarios are thought provoking as they offer perspectives on how to rethink tourism.

Chapter 8 presents a set of future scenarios for the rural province of Friesland, the Netherlands. The case study, commissioned by the provincial administration, addresses regional development in general, and, therefore, is not tourism specific, however, it does show the huge implications for the future of tourism and highlights the importance of taking a multi-domain, integral approach to the development of tourism and regions at large. The project was conducted in 2017–2018.

The Fryslân Futures Lab 2030 is the subject of Chapter 9. The purpose of the project, commissioned by the provincial administration of Friesland and started late 2019, is to create a story/image for the province in 2030 that serves as an inspirational starting point for multiple long-term strategies and for acquiring the competence of strategic foresight. It should satisfy the need to create a dialogue among the policymakers within the provincial government of Friesland, across various domains. It should inspire and challenge the provincial government and stimulate creativity.

Chapter 10 explores scenarios for inbound tourism to the Netherlands. The purpose of the project, conducted in 2018–2019 and commissioned by NBTC Holland Marketing, was to develop a number of future scenarios that would enable NBTC Holland Marketing to make estimations about the size of inbound tourism until 2030. The chapter presents a baseline scenario and four alternative futures scenarios.

In Chapter 11 the future of the Notting Hill Carnival is explored. The project was conducted in 2013, a year before the 50th anniversary of the Carnival. Its main aim was to holistically engage the Notting Hill Carnival community in collaboratively planning for its future, alongside academics and other individuals who could contribute meaningfully to the planning process. The project intended to provide inspiration for the cultural and entrepreneurial innovations which would sustain the Notting Hill Carnival until 2020 and beyond. The project was commissioned and funded by King's Cultural Institute's Creative Funding Programme.

In response to the growing agitation of residents in European cities towards tourism between 2015 and 2019, a study was conducted by the centre of Expertise in Leisure, Tourism and Hospitality, of which the scenario part is presented in Chapter 12. The purpose of the study was to get a better understanding of visitor pressure ('overtourism') as perceived by the cities' residents and experts, and to explore the future development of urban tourism by means of scenarios. The study was conducted in and with the Destination Management Organisations (DMOs) of major European cities such as Barcelona, Lisbon, Berlin, Copenhagen, Amsterdam, Munich, Mechelen, Ghent, Antwerp, Bruges, Leuven, Tallinn and Salzburg. Its outcomes were published in collaboration with the World Tourism Organisation.

The final chapter provides a critical reflection (Chapter 13). It evaluates the previous chapters and identifies the contributions from an ontological, epistemological and methodological perspective. In addition, the chapter identifies a series of issues and emerging trends in scenario-planning research and practice that will shape the future debate about the topic.

1.5 Discussion Questions

(1) Which global developments, different than the ones mentioned in the chapter, would have a long-term effect on international tourism?
(2) What would be the possible consequences for international tourism that you are thinking of? For tourists? For destinations? For businesses? To facilitate the discussion, you could choose a specific destination with its visitors and businesses as a starting point.
(3) What could the industry do now to anticipate and take advantage of such consequences?

Notes

(1) Internationale Tourismusbörse, in English: International Tourism Fair.
(2) https://www.itb.com/en/.

2 Is Scenario Planning a Theory Builder or a Research Methodology?

Learning Points
- Scenario planning is the dominant foresight methodology, but why?
- Scenario planning as a theory builder.
- Scenario planning as a learning framework.

2.1 Introduction

When practiced, does scenario planning have a theoretical proposition? The literature on scenario planning reports that it is a practice driven methodology, shaped by practitioners for practitioners (Amer *et al.*, 2013). However, practice and theory are closely interlinked (Mingers & Rosenhead, 2004; Yeoman, 2004), so could the practices of scenario planning, in fact, be its theory? This chapter provides an overview of future studies and then explores the literature on scenario planning, the most prominent foresight methodology. The chapter then delves into the success of scenario planning and asks whether scenario planning is a research methodology or theory builder.

2.2 What is Future Studies?

The use of 'future' in the English language dates back to the 14th century. It derives from the Latin *futurus*, meaning 'about to be', which became assimilated to French as 'futur'. Broadly speaking, 'future' and its translation refers to the time that is to be or come hereafter. It is not clearly delineated in terms of time horizon; it may mean tomorrow, next year, the coming decade, the next 20, 30 or 50 years or even forthcoming centuries (Asselt *et al.*, 2010). Scientific research, which is positivist in nature, is grounded in empirical research that is objective and data rich. However, data about the futures cannot be gathered from surveys or in a positivist way, as the future hasn't occurred yet. However, we live in a society which is data driven and objective (Yeoman & Postma, 2014).

The field of future studies has a theoretical grounding in the academic journals of *Futures, Foresight, Long Range Planning* and *Technological Forecasting and Social Change*. Futurists, those researchers who practice the future, may have completed a Masters or PhD degree in Future Studies at Houston University (USA), University of Turku (Finland) or Swinburne University (Australia). Governments, international agencies and large corporations have, since the 1960s, being interested in future and futures methods. Famous examples included *The Year 2000* (Kahn & Wiener, 1967) and the Club of Rome's *Limits to Growth* report (Meadows *et al.*, 1972). In more recent times, global businesses, such as Shell (www.shell.com/scenarios) are examples of scenarios developed in a business context or the World Economic Forum's (https://www.weforum.org/) focus on the future.

Humankind did not always contemplate the future as a realm of action (Adams & Groves, 2007). In early times, the future was considered a sacred domain ruled by the gods. Only in modern times did the idea arise that humans could influence or even shape the future. This view of the future, as a realm of action, encouraged interest in contemplating the futures. The future can be engaged in a variety of ways from utopia novels, such as dystopian novels (Wells, 1902, 1977), science fiction (Yeoman & Mars, 2012) or a business style report (https://www.foresightfactory.co/). Economists deploy econometrics to forecast the future economy. Climate models are used for a variety of purposes, from the study of dynamic weather and climate systems to projections of future climate. Demographers use vital statistics that track births and deaths, combined with data, such as marriage, divorce and migration, to forecast populations. This approach to the future is predictive, numerate and singular. Here, the future is strongly linked to the past, connecting the past to the future (Yeoman & Mars, 2012). Whereas others believe in multiple futures as accuracy cannot be achieved with a singular, predictive approach to the future (Yeoman & McMahon-Beattie, 2005, 2014; Yeoman & Postma, 2014).

2.3 What are Foresight Methods?

Foresight is a human capacity which allows people to think ahead, mode, create and respond to future eventualities (Conway, 2006). Foresight methods are methodologies and processes used within future studies to understand the future (Karlsen & Karlsen, 2013). These tools range from forecasting methods, including econometrics (Munro & Yeoman, 2005) and predictive consumer behaviour models using structures equation modelling (Luo, 2002) or complex modelling techniques used for climate change scenarios (Raäisaänen, 2007), to more qualitative methods, such as systems thinking (Postma & Yeoman, 2021) and cognitive mapping (Farsari *et al.*, 2011; Yeoman *et al.*, 2005b).

The most common methods are environmental scanning (Platform, 2022), trends analysis (Yeoman *et al.*, 2022) and Delphi panels (Linstone & Turoff, 2011). However, the dominant method is scenario planning (Amer *et al.*, 2013; Martelli, 2014a).

2.4 What is Scenario Planning (or Scenarios)?

Scenario planning is a popular approach for addressing uncertainty in strategic decision making. An open and adaptable approach from its inception, scenario planning has developed into separate schools and is now used across a wide range of research fields and practical settings (López-Rodríguez *et al.*, 2023). Scenarios are the future states or a representation of the future, whereas scenario planning is the methodology to create the scenarios. Both terms, are used interchangeably in the literature and mean the same thing (López-Rodríguez *et al.*, 2023)

Scenario planning is used to create awareness and prepare for uncertain future scenarios and is used to deal with the inherent uncertainty in short-term disruptions (such as, for example, flooding, COVID-19, terrorist attacks or a financial crisis), for exploring long-term developments (e.g. climate change scenarios or social responses to public health interventions) and help to test the robustness of different strategies against multiple possible futures. The use of scenario planning is reported in the literature under different names: scenario development, scenario planning and scenario thinking (Oner, 2010; Varum & Melo, 2010), or referring to a specific approach to constructing scenarios, like participatory scenario planning (Poskitt *et al.*, 2021).

So, what is scenario planning? A number of studies have set out to define scenario planning (Amer *et al.*, 2013; López-Rodríguez *et al.*, 2023) but they all overlap to a certain extent. Some authors reflect scenarios or scenario planning as an approach or method, others as a technique or tool (Bishop *et al.*, 2007). However, what they all have in common is that scenarios planning is constructed, not with the goal to predict the future or make a forecast, but rather to construct multiple possible stories of the future situation. Scenario planning includes a time horizon from 3 to 50 years plus. Authors emphasise the future with uncertainty, innovation, change and complexity. Different techniques or tools are suggested to deal with the uncertainty while constructing scenarios, such as sensitivity analysis, matrix score, etc. Scenario planning uses stories to portray the future and is usually workshop based in which participants use activities to construct different scenarios. Right at the centre of the scenario planning process is the scenario 2 × 2 matrix in which four scenarios are presented based upon two uncertainties (Ramirez & Wilkinson, 2014a). A workshop process ensures that scenarios are used for discussion, reflection, strategies and decisions. The scenario planning process is often described as a participatory learning process in which the emphasis is placed on scenario thinking, in order to create change (Postma & Yeoman, 2016, 2021; Wilson, 2000). Schoemaker (1993) noted that three prime characteristics set the scenario approach apart from the then traditional planning tools: (1) it is an approach centred on a script or narrative; (2) it places uncertainty across, rather than within, individual models and (3) it portions out complex future possibilities into discrete states that are easier to assess, use and compare.

2.5 Scenario Planning: Theory or Methodology?

Research by a number of authors (Cordova-Pozo & Rouwette, 2023; López-Rodríguez *et al.*, 2023; Wade, 2021) identifies that there has been considerable growth in the literature on scenario planning in the past decade. The literature has been organised around a number of schools, themes and techniques. Clusters of authors have approached the virtues of scenarios planning in different ways. For example, Bradfield *et al.* (2005) focuses on the development of scenario techniques, while Schnaars (1987) identifies that a significant part of the literature describes and explains how to carry out scenario planning, concluding that the literature falls into two parts. The first of these is how scenario planning is carried out by large companies that offer reflective based advice on the process of scenario planning. The articles by Pierre Wack are examples of this writing (Chermack & Coons, 2015). Bradfield *et al.* (2005), Chermack (2005) and Keough and Shanahan (2008) have reviewed various methodological approaches and guidelines presented in the literature. Bishop *et al.* (2007) studied more than a dozen techniques for planning and commented on utility value, strengths and weaknesses of these methodologies. In the second part, the process of scenario building (Durance & Godet, 2010) often cite the literature of Schwartz (1996) and Martelli (2014b). Schwartz describes in detail each step of the scenario-building process, plotting drivers to development various scenarios. Martelli (2014b) offers a unique, careful insight on the process of action learning, focusing on the relationship between organisation and the environment. Schoemaker (1993) presents a very comprehensive and detailed account of the 10 steps required in scenario building. Developing scenario planning as a theoretical base has been advocated by Chermack (2005) using Dubin (1978), based upon a number of law's of interaction, organisational learning, mental models, boundaries and system states. However, the essence of Chermack's approach is the improvement of the methodology, creating rules and formulating laws. This is still about improving scenario planning as a methodology or as a means of investigation.

Although the writings about scenario planning are fundamentally a methodological contribution, there has been a renewal to develop scenario planning from an empirical and theoretical base. Publications about scenario planning can be found in dedicated journals like *Futures, Foresight, Technological and Social Changes* and the *Journal of Future Studies*. Scenario planning is prevalent in many management journals like the *Journal of Technology Management, Harvard Business Review, Sloane Management Review* or *Long Range Planning* (Varum & Melo, 2010). We have also seen the emergence of specialist journals that explore the futures of specific fields (i.e. *Journal of Tourism Futures* (Yeoman *et al*., 2015a)). These journals, which are theoretically and empirically focused, have facilitated the moving of scenario planning from a methodological chaos (Varum & Melo, 2010) or fog (Asselt *et al*., 2010).

Martelli (2014b) argues that scenario building is within a framework of plurality, uncertainty, time and complexity. These are epistemological arguments about the structure of knowledge (Mingers, 2014; Mingers & Brocklesby, 1997; Mingers & Rosenhead, 2004). This epistemological argumentation takes scenario planning into paradigm classification and the future studies. Amara's (1974) typology of probable, possible and preferable is goal orientated and conceptualises pathways of multiple futures, which is replicated in many theories of the future. In Dator's (2009) *Alternative Futures at the Manoa School*, the first law states that 'the future cannot be predicted but alternative futures can be pondered' (2009: 9). The second law states that 'any useful idea about the futures should appear to be ridiculous' and 'we shape our tools and thereafter our tools shape us' (2009: 449). Sardar's (2010) laws of futures studies are based on the notions of wicked problems, diversity, scepticism and the present. Sardar and Dator's laws overlap in the notion that there is no one future, thus multiple futures prevail. All futures must include a degree of ridicule and scepticism in order to go beyond the present. In addition, there is no point in preparing and actioning the future if there is no relevance to the present. Habermas's (1974) typology of interests of knowledge was amplified by many authors in the 1970s and 1980s, including Sandberg's (1978) typology of technical, hermeneutic and emancipatory futures, and then later further developed by Slaughter (1996). Masini's (1989, 2006) extrapolation, utopia and vision approaches were a synthesis of Amara's (1981) and Mannermaa's (1991) theories. Inayatullah's (2010) epistemological approach to futures studies was based on the predictive-empirical, cultural-interpretative and critical-post-structuralist approaches. Hideg (2002, 2013) presents three core paradigms of futures studies, positivism, evolutionary and critical futures arguing that there is competition between paradigms, whereas Slaughter (2002) advocates that futures studies align with critical realism because of human actors, power and context. Hideg (2013) identifies a number of delimits or blind spots of the paradigms which make a substantive contribution to theory through integration with practice.

The above review leads us to a simple question, what is 'theory' and what is 'methodology'? Theory, a common academic term, 'comes from the Greek' (Smith & Lee, 2010: 28):

> The term traditionally denotes contemplation, speculation or a world view. Other definitions include a mental scheme or course of action for doing something, a systematic statement of facts or principle on which a body of knowledge is founded, abstract knowledge or speculation. A frequent academic connotation of the term is that what is presented as theory reflects intellectual sophistication and is, therefore, superior to the atheoretical. (Smith & Lee, 2010: 28)

Of more direct relevance, is the assertion that theory is a body of logically interconnected propositions (that) provides an interpretive basis for understanding phenomena (Smith & Lee, 2010). Hierarchically, traditional theory is in the form of traditional theory found in natural science, then there are theories synonymous with a priori and empirical studies. Lower order theories relate to epistemological classifications and untested assumptions. Bates and Tucker (2010) argue that theory is a system of principles and relationships posited to explain a specific set of assumptions, whereas methodology is a set of methods developed according to a paradigm about how best to research and learn about a natural or social phenome. Theories and frameworks of futures studies are epistemologically constructed, thus representing the assumptions, characteristics and forms of knowledge associated with a particular paradigm.

2.6 Methodological Domination

Future studies research is dominated by a commitment to research methods almost as an end in itself, with scenario planning being the default methodology (Asselt *et al.*, 2010). This is reinforced by the first issue of *Futures* in 1969, futures were about method.

> Futures (concepts has arisen because of the need for systematic methods of dealing with the enormous number variables that must be taken into account when forecasting (Hales, 1969: 2)

According to Karlsen *et al.* (2010) there has been a general failure to examine and explicate the relationship between method and success. What is missing, beyond epistemological arguments (Börjeson *et al.*, 2006; Chermack & van Der Merwe, 2003; Derbyshire, 2017; Mackay & Tambeau, 2013; Marchais-Roubelat & Roubelat, 2008; Martelli, 2014b; Powell, 2001; Thompson, 2011; Yeoman & McMahon-Beattie, 2018b) is why scenario planning is successful.

According to Amer *et al.* (2013: 23):

> Scenarios are considered a valuable tool that helps organisations to prepare for possible eventualities and makes them more flexible and innovative. Scenarios are outlines of some aspects of future and generally scenario refers to an outline of the plot of a dramatic work, script of a motion picture or a television programme.

However, as Varum and Melo (2010: 365) argue:

> At empirical level there is a notable lack of research on the use and effects of scenario planning in business. There is no empirical data in the literature which documents the popularity of scenario techniques from the early 1980s to the present day. Similarly, there is a lack of extensive studies on the effects of scenario planning on company performance and competitiveness.
>
> The importance of an accurate assessment of the business environment for the development of a corporate strategy goes unchallenged. For managers this article raises awareness with regard to future analytical methods, and in particular, to the advantages (and disadvantages) of using scenario planning and its potential contribution to the competitiveness of firms.

Thus, scenario planning is a valuable tool in strategic planning in business, but its effectiveness and success in not understood, as per Varum and Melo (2010). According to Asselt *et al.* (2010), their simplicity, utility value and ability to make sense of the future through participation are the keys to their success.

2.7 Theory Building

Theory is defined as 'a set of interrelated constructs (concepts), definitions and propositions that present a systematic view of phenomena by specifying relations among variables, with the purpose of explaining and predicting the phenomena' (Kerlinger & Lee, 2000: 11). Theory building is the ongoing process of producing, confirming, applying and adapting theory (Lynham, 2000). Within the literature, scenario planning approaches claim a theoretical contribution, from adapting theory from various outside fields untethered to the ontological, epistemological and methodological assumptions of futures studies to name but a few, consider the theory of theory of aesthetics (Ramírez & Ravetz, 2011), the behavioural theory of the firm (Gavetti, 2012), complexity theory (Wilkinson *et al.*, 2013), empirical philosophy (Rowland & Spaniol, 2017), managerial cognition (Hodgkinson & Clarke, 2007), evolutionary theory (Evans, 2011), organisational learning (Chermack & Walton, 2006), ontology (Bergman *et al.*, 2010), potential surprise theory (Derbyshire, 2017), sensemaking (Blass, 2003; Klein *et al.*, 2006; Moriarty, 2012), consensus and social negotiation (Rowland

& Spaniol, 2017). From this vantage point, there is an appreciable variety of theory in the scenario planning literature.

As Derbyshire (2016: 1) points out:

> Despite this, it is still widely held, including by those having carried out what theoretical work does exist, that scenario planning remains under-developed theoretically.

As Spaniol and Rowland (2018: 36) conclude:

> All these contributions to theory appear as though they help fill the proverbial 'hole;' however, as they mount, they paradoxically clutter the literature and thereby distance the field from the sort of shared foundational theory indicative of paradigmatic 'normal science'.

2.8 Scenario Planning as a Theoretical Framework

The key purpose of scenario planning is to change/alter mental models through dialogue, conversation and decision making, as a participatory group process, thus, contributing to learning and an increased capacity to think in innovative and challenging ways (Bradfield, 2008). Scenario planning theory places strong emphasis on revealing and reconstructing mental models. Theorising assumes that organisations are systems of feedback loops that spread the dominant mental models and cultural artefacts through interaction (Huff & Jenkins, 2002; Weick & Roberts, 1993; Yeoman, 2004).

The framework approach draws upon Pearce's (2012) proposition of tourism theory, building through conceptual and theoretical frameworks, in which Pearce illustrates scenario planning as an appropriate framework. At the heart of the scenario planning framework is the 2 × 2 matrix allowing researchers to conceptualise a particular phenomenon related to the future. Chermack (2007) argues that scenario planning can be considered as a mode of theory building. This argument is based upon the fact that scenario construction and theory building share several similar key characteristics in terms of purpose, process and outcomes. Both scenario planning and theory building consider theories-in-use mental cognitive maps. They are described as 'disciplined imagination' processes to formulate multiple plausible solutions to a problem. Scenario planning develops different plausible futures, akin to generate different theories about the future. Each scenario can be considered a theory about the future once developed in detail (Chermack, 2007).

Scenario planning and theory building have similarities and differences in the steps of the processes. But the notable difference lies in the data analysis process. Theory building relies on well-documented processes, such as statistical analysis, case study research, grounded theory, meta-analysis or comparative analysis towards statements of truth about the phenomena. This refers to the explanatory claim developed by Bergman *et al.* (2010) and Yeoman and Postma (2014) that underlines scenario construction as a form of theory building.

Fergnani and Chermack (2021: 1) argue that 'weak theoretical foundations prevent the field (futures and foresight studies) from becoming a recognised academic discipline of study in the academic establishment'. One of the issues is that futures researchers lack training in theory building and theory testing, hence, the focus on methodology and not recognised theory (Yeoman & McMahon-Beatte, 2018b). Getting familiar with theory building and theory testing is a way to enhance the theoretical development of this field.

The father figure of scenario planning is without doubt James Dator from the Hawaii University System. Dator's (2009) *Alternative Futures* is the most cited paper in scenario planning, foresight and future studies and is considered a theoretical framework and theory builder. Dator (2015) focuses on plurality of futures, presenting four generic alternative futures of Grow (Continued Growth), Collapse (New Beginnings), Discipline (Sustainable) and Transformation, which are constructed from seven drivers ('driving forces').

Leading Finnish futurist, Professor Anne Bergman, in Bergman *et al.* (2010), considered an ontological philosophy which had similarities to Dator's (2009) *Alternative Futures*.

Bergman's ontological classification was on the two dimensions of 'truth claims' and 'explanatory claims'. The term 'truth claims' refers to the 'beliefs' on what is going to happen in the future; and the term 'explanatory claims' means the authors' statements that explicitly indicate mechanisms as causes behind the events or states forecasted, thus, an ontological position. The advantage of the Bergman *et al.* (2010) ontology is their classification of futures studies helps us to understand why there are different possible futures (Bergman *et al.*, 2010) and stretches the understanding of the future (Yeoman & McMahon-Beattie, 2016). The weakness in their work is the language, that is truth and explanation are usually used with an epistemology perspective.

Bergman *et al.* (2010) propose two ontological dimensions with two values on each resulting in four possible outcomes: forecasts with truth claims that indicate mechanisms (predictions); forecasts with truth claims without indicating mechanisms (prognosis); forecasts without truth claims but indicating mechanisms (science fiction); and forecasts without truth claims and mechanisms (utopias/dystopias), see Table 2.1. This future classification provides separate and differentiated scenarios in the range of Prediction, Prognosis, Science Fiction and Utopia/Dystopia.

Yeoman and McMahon-Beatte (2018b) acknowledge the limitations of scenario planning as being too focused on logic, in particular the 2 × 2 matrix tends to produce scenarios focused on logic and rationality.

Table 2.1 Types of forecasts

	No Explanatory Claims	*Explanatory Claims*
Truth Claims	*Prognosis:* An alternative, feasible, sophisticated, answers, complex, expert-based opinion, trust.	*Prediction:* Near future, precision, scientific ambition, highly probable, certain, historical.
No Truth Claims	*Dystopia–Utopia:* Visions, no places, crisis, journeys, warning.	*Science Fiction:* Impossible, implausible, unclear to some, sceptism, fictional narratives, speculative, fiction narrative.

Source: Bergman et al. (2010).

This is what Gordon (2020) observes as matrices being too normative rather than adaptive in presenting the future. Ramirez and Wilkinson (2014b) argue that the scenario matrix is enshrined in plausible and alternative futures. In particular, the matrix is shaped by the philosophy of logic which treat's all scenarios within a scenario set (or matrix) as are equally probable through discourse of 'coherence', 'plausibility', 'internal consistency', 'equal probability' and 'logical underpinning' (Bradfield, 2008). Extending this logic approach through the language of plausibility is Ramirez and Wilkinson (2014b) writings on intuitive logic, where plausibility becomes the centre of discussion derived from historical determinism. Plausibility emerges from and supports strategic conversations that question whether that which has been impossible might become possible and which investigate how that which has been possible might end. The weakness of this approach is that it hinders transformational thinking, unthinkable and creativity. This approach is too deductive as an outcome.

As Ramirez and Wilkinson (2014b: 258) note:

> Drawbacks of the deductive building method can be that the resulting scenario set is considered too obvious or simplistic; for example, assuming independence between variables rather than exploring more interesting challenges implied by their co-evolution. Thus, one choice of one axis might be the emphasis on policy making versus environment, thus removing the future possibility space in which there a progressive role for markets and enterprise in achieving environmental improvements.

Thus, it is necessary to apply an ontological framework, such as Bergman's (2010) types of forecasts or Dator's (2009) *Alternative Futures* as an overlay on scenario matrices too overcome the boundaries of logic and stretch our thinking and understanding of the future.

2.9 Scenario Planning as a Learning Framework

As Pierre Wack (1984: 258) notes:

> Good scenarios are not enough. To be effective, they must involve management, top and middle, in understanding and anticipating the unfolding business environment much more intimately than would be the case in

the traditional planning process. Scenarios can be successful in structuring uncertainty only when (1) they are based on a sound analysis of reality and (2) they change the decision maker's assumptions about how the world works and compel him to change his image of reality. A willingness to face uncertainty and understand the forces driving it requires an almost revolutionary transformation in a large organisation. And this transformation process is as important as the development of the scenarios themselves.

Wack used the words 'corporate microcosm' to refer to the set of assumptions (mental models) in the minds of management that they used as the basis for decision making. These perceptions were where strategy came from and, therefore, influencing these assumptions in a constructed process of learning and adjusting these assumptions was the site of highest leverage for dealing more effectively with uncertainty in the environment based on the assumption that the end result of scenario planning is not about a more accurate picture of tomorrow but better decisions about the future (Chermack, 2004; Chermack & van Der Merwe, 2003). Thus, this raises the question about whether scenario planning can be used to create better decisions. The answer to this question lies in scenario planning as a learning process. As Chermack points out, this perspective has four criteria:

(1) the individual construction of knowledge;
(2) social influences on individual constructions;
(3) the situatedness and contextual requirements of knowledge construction; and
(4) the social construction of reality.

Therefore, scenario planning becomes a framework for learning to change mental models through participation, where the emphasis from an operational perspective is one participation learning (Yeoman, 2012b) through workshops, interaction and group discussion. The importance of comparative learning through scenarios and visual learning is to see impact, connectivity and the whole picture (Gangwer, 2009). An important feature of the scenario planning process is thinking, in which workshop activity allows participants to break down complex issues, layer issues and understand the parts of the problems, rather than being overwhelmed with complexity. In particular, workshop participants go through the process of zone of proximal development or scaffolding (Holton & Clarke, 2006; Piaget & Brown, 1985) in order to breakdown, structure and learn about problems.

2.10 The ETFI Method as a Framework of Theory and Practice

The key purpose of scenario planning is to change mental models through dialogue, conversation and decision making as a participatory

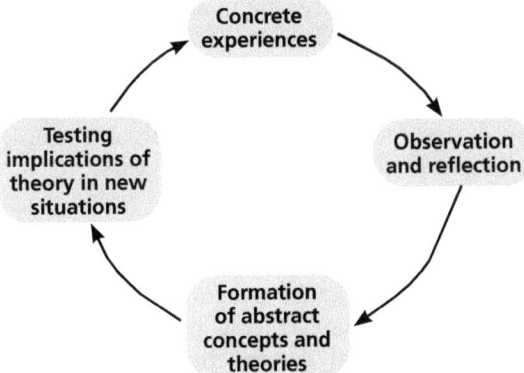

Figure 2.1 Learning loop

group process, thus contributing to learning and an increased capacity to think in innovative and challenging ways. It is fundamentally a model of practice with theory building and educational foundation.

The scenario planning process follows a structured process, which is iterative, systemic and systematic. The ETFI method (elaborated upon in subsequent chapters) embodies theory and practice and is fundamentally a model of continuous learning (Van der Heijden, 2005; Van der Heijden et al., 2002), which incorporates Kolb (1984) learning loop (see Figure 2.1) advocated by Piaget and Brown (1985), in order to change the mental models of stakeholders, actors and decision makers.

Learning is based upon the process of understanding the problem through a learning experience, reflection and raising awareness to bring about new patterns and thoughts previously not surfaced. Surfacing is about shifting mental models in order to shift decision making and actions.

2.10.1 Communities of practice

Lave and Wenger (1991) discuss the idea of communities of practice, which is about bringing together persons associated with problems and those who would benefit from solving the problems, from stakeholders to actors, over a period of time. Communities of practice are not about a one-off situation but a continued process of share understandings and benefits aimed at shifting learning and mental models. This involves practices, social structures, power relations and conditions for legitimacy, among other characteristics. Chermack and van Der Merwe (2003: 253) acknowledge that scenario planning deals heavily with differing communities of practice. It is also what Van der Heijden et al. (2002) call scenario thinking rather than practice. It is about creating a collective mindset of change.

Thus, the process of scenario planning creates categories for stakeholders, competitors and, in order to accomplish the task of shifting mental models, must frequently consider the perspectives of these difference communities of practice. Furthermore, scenario planning is often a process that includes joining several communities of practice through the telling of plausible stories of the future.

2.10.2 Social reality

Aligned with communities of practice is the social construction of reality – that is, reality constructed by society and is constructed socially. This links to social constructionism in which participants construct their reality around them, both objectivity and subjectivity. Berger and Luckmann (1967) conclude that objective reality relates to the process of becoming human taking place in an interrelationship with an environment, whereas subject reality is the socialisation of knowledge and the interaction with others. This is at the heart of learning, especially through workshops of communities of practice. The process of social reality brings about strategic conversation around the problems or scenarios being worked upon. Scenarios use a language of options, plausibility, decisions, actions and create a reality through sense making. An important feature of the social reality is the facilitator, who in many cases is the scenario planner (Yeoman & McMahon-Beattie, 2005).

2.10.3 Facilitation

The concept of facilitation is being used increasingly in management literature (Eden & Ackerman, 1998) and the importance the scenario planning literature (Amer et al., 2013; Chermack et al., 2015; Derbyshire, 2017; López-Rodríguez et al., 2023; Rowland & Spaniol, 2017) is frequently mentioned, with many writers advocating an explanation of the concept and its meaning within scenario planning. Bentley (1994) states the word 'facilitate' comes from the Latin 'facilis' which means 'to make easy', meaning to 'active process, means of facilitating or moving forward' (Yeoman, 2004: 36). This is the role of the facilitator who is often the scenario planner. The scenario planner manages the process of learning, group dynamics and scenario planning process. He or she is a 'Jack of all trades or a professional "do-it-yourself" person' (Levi-Strauss, 1966: 17).

The scenario planner in action is a person who pieces together the activities into a coherent set of scenarios that make sense of the complexity of data and phenomena becoming the interpreter of reality. It is the scenario planner in action that deals with politics, reality, time pressures, individuals and group processes. Reality takes us back to the concept of pragmatism is finding a satisfactory outcome rather an optimal outcome, just satisficing rather than optimising (Ackoff, 1978).

2.11 Concluding Remarks

In order to understand the future, it is important to structure and make sense of it. Scenario planning is the most dominant and most practiced foresight methodology because of its simplicity in dealing with complexity. The history of scenario planning is embedded in practitioner development through pragmatism and adaptation. Only most recently have thoughts turned to the theoretical underpinning of scenario planning as a theory builder based upon the ideas of plurality, in order to represent the future. The cornerstone of futures studies is plurality, as no one can predict the future, but we can discuss a series of alternatives. Here, it is about stretching our understanding of the future from an ontological and epistemological perspective based upon Dator's (2009) *Alternative Futures* and Bergman's forecast as a means of improving the 2 × 2 scenario matrix as a theory builder.

We have stated that the key purpose of scenario planning is to alter the mental models of thought through dialogue, conversation and decision making as a participatory group process, thus contributing to learning and an increased capacity to think in innovative and challenging ways (Bradfield, 2008). This is the focus of the ETFI scenario planning model. Scenario planning is embedded in practice and its success is because practitioners have adapted the process due to the reality of pragmatism. Throughout the remainder of this book, you will find how the ETFI model of scenario planning works, based upon these theoretical underpinnings and how ETFI has adapted through scenario planning interventions and projects.

2.12 Discussion Questions

(1) In your own opinion, is scenario planning more about theory or practice?
(2) Reflecting upon this chapter, what are important concepts that are associated with successful scenario planning? Elaborate your answer with reasoning.
(3) How do you ensure plurality and difference when creating scenarios?

3 Strengthening Resilience in Tourism through Scenario Planning

Learning Points

- Resilience is becoming a key feature of destinations, in order to survive and adapt in times of change.
- Bridging theories on resilience and scenario planning contributes significantly to a shift in focus from reactively fixing problems to proactively addressing the structural issue of adaptive capacity building for the sake of resilient destinations.
- When we bring the concept of resilience to a specific tourism context, we should be explicit about the *resilience of what, resilience to what* and *resilience by whom*.
- Destination resilience is not a given, it is a complex process that requires much resource investment, and benefits greatly from scenario planning and strategic foresight.

3.1 Why Destination Resilience?

Resilience is generally understood as '*the capacity of a system to absorb disturbance and reorganise while undergoing change to still retain essentially the same function, structure, identity and feedbacks*' (Walker *et al.*, 2004: 1). The concept of resilience and the theoretical discussions regarding the concept can offer helpful insights to further improve the development, management and governance of tourism destinations. Also, there are various links to be made between resilience, on the one hand, and scenario planning and strategic foresight, on the other hand. For instance, scenario planning can help to create important feed-forward loops (predicting, planning) that support actors to understand the impacts of forces driving change and explore possible future states to which destinations might evolve, in order to develop courses of action, if necessary. Hence, scenario planning and strategic foresight can contribute significantly to the desirable shift in focus from reactively fixing continually emerging problems to proactively addressing the structural issue of adaptive capacity building for the sake of resilient destinations.

In this chapter, we explore the value of resilience, taking the standpoint that tourism destinations are open, complex and (potentially) adaptive systems. These arguments are developed in full detail (e.g. in Baggio, 2008; Brouder & Eriksson, 2013; Ma & Hassink, 2013; Hartman, 2016, 2018a, 2021; Hartman et al., 2020). When we bring the concept of resilience to a tourism context, we believe it is important to be explicit about the *resilience of what, resilience to what* and *resilience by whom* (Sellberg et al., 2015). The following sections dive deeper into these questions. We end the chapter with concluding remarks on the linkages and the ways in which theories of resilience and scenario planning methodologies can reinforce one another.

3.2 Resilience of What?

The *resilience of what* in this book concerns tourism destinations. Destination are generally rather complex entities that consist of many interrelated products, sectors and institutions and their mutual interactions (Ma & Hassink, 2013). More specifically, in the words of Brouder and Eriksson (2013: 373) tourism destinations can be regarded as a *'bundle of many sources of evolutionary change'* that is driven by *'multiple levels of agent interaction in the form of labour, firms, networks, technologies and institutions'*. The resilience of tourism destination then depends on many different parts that together influence and determine the extent to which the capacity to adapt is suitable and sufficient to manage changing circumstances. Alternatively stated, destinations are always in a process of responding to and anticipating both endogenous, as well as exogenous shocks and stresses that influence its development and the development of agents or actors within these systems (Hartman, 2018b). Many forces that drive change take place rather autonomously. And since tourism destinations conceptualise as open, complex adaptive systems and consist of many constituent parts, there is no single actor in complete control and the ability to fully command and control such forces is impossible. Simultaneously, this implies that adapting to impacts and consequences of forces that drive change is (becoming) a crucial capacity. The capacity to adapt, and therefore the resilience of a tourism destination, depends strongly on its constituent parts, for example, the individuals, firms, organisations, branches, clusters and their connectivity and interactions. This is discussed in more detail in Section 3.4 *Resilience by Whom?*

3.3 Resilience to What?

The development of the tourism industry, and tourism destination in general, is strongly steered and shaped by all sorts of processes that relate to the dynamics of today's society. Demographic changes have their effect on the volumes and types of visitors. Economic changes may influence welfare levels and buying behaviour. Environmental changes relate to the climate, sea level and biodiversity can become drivers of change for the tourism industry. (Geo)Political changes may result in the opening up of areas to visitors or the closing off of places to visitors. Technological changes related to virtual reality, zero-emission mobility and digitalisation affect how visitors travel and behave on site and require businesses in tourism to adapt. These are just a few of the possible forces that drive change. More in general, the tourism industry and tourism destination at large need to be able to manage change, crisis, shocks, stresses, disruptions, perturbation which also include *'individual traumas, terrorist attacks, natural disasters, natural developments like global warming, global economic crises, major plant closures, technologies becoming obsolete, the fall of complete industries, political transformations…and so forth'* (Boschma, 2015: 734).

All in all, the tourism industry needs to be able to manage a potentially wide variety of shocks ('fast variables') and stresses ('slow variables'), as discussed in Lew (2014). Slow variables that destinations need to anticipate include, among others, climate change, developments in the (macro)economy and demography and increasing flows of tourists. Fast variables include, for instance, changing weather conditions, natural disasters, geopolitical events, terrorism, changes in consumer preferences and new technologies. Such variables (see Table 3.1 for an overview) need to be included in scenario planning processes, as their impact can range between being weak signals of change to major forces driving change.

Table 3.1 Shocks and stresses

Natural and environmental disasters	*Tsunami, earthquake, volcanic eruption, fire, wildfire.*
Climatic risk and extreme weather	*Flood, storm, hurricane, tornado, thunderstorm, cyclone, typhoon, shortage of snow, coastal erosion, landslide and debris flow, drought, extreme heat, extreme cold and winter storm, blizzard, avalanche, etc.*
Sanitary disaster and epidemics	*Severe Acute Respiratory Syndrome (SARS), dengue, zika, Ebola virus, bird flu, COVID-19.*
Industrial risks	*Nuclear disaster, chemical disasters, oil spills toxic and industrial pollution.*
Political and social crises	*Arab Spring, revolutions, war, attacks, terrorism, strikes, crime, security.*
Macroeconomic shocks	*Financial crises, economic instability, price of oil, inflation.*
Population	*Ageing population, middle class development.*
Urbanisation	*Excess urbanisation.*
Technology	*Digital disruptions.*
Economy	*Demand transformation, demand slump, decline of arrivals.*
Ecology	*Greenhouse effect, biodiversity.*

Source: Based on Fabry and Zeghni (2019).

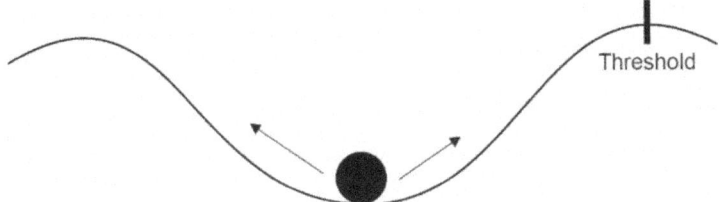

Figure 3.1 Out of balance

In the context of shocks and stresses, as explained in Hartman (2018b: 67), the challenge is '*to become robust enough to endure perturbations and flexible enough to recover or to re-develop/re-invent itself*'. In other academic contributions on resilience, the series of figures presented below tend to reoccur. This set of figures help to explain resilience. Figure 3.1 shows a situation in which shocks and stresses bring a destination in an out-of-balance situation, also known as 'out-of-equilibrium' in the literature on complexity theories (De Roo, 2012).

When such situation occurs, and we oversimplify here for explanatory reasons, two general types of efforts can be taken. First, efforts can be taken to re-balance the destination, whereby actors aim to bring the destination and its development back to the situation where it was before for the impacts of the shocks and/or stresses (see Figure 3.2). Second, efforts can be taken to seek a tipping point that functions as a threshold to transition towards fundamentally different types of structures, functions, organisations and identities. For the tourism industry, this could be coupled with new business models or new cooperation partners. For destination development and management, this could include fundamentally different goals (e.g. value over volume or regeneration over exploitation), new stakeholders, different organisational structures and institutional frameworks (laws, rules, regulations, etc.).

Potentially, the impacts of shocks and stresses leave a destination no other option than to drastically change when the forces that drive change push the destination so far out of balance that re-balancing is impossible, or undesirable, and alternative future states should be sought. An illustration of such situation is provided in the New York

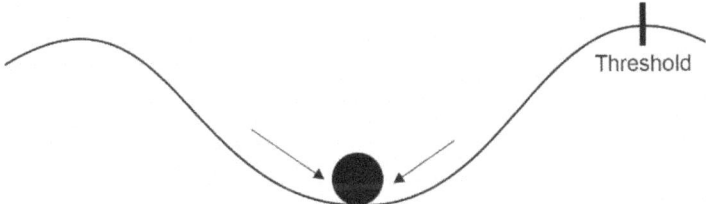

Figure 3.2 Re-balance

Times (31 March 2021 'In Empty Amsterdam, Reconsidering Tourism') citing Geerte Udo, the director of amsterdam&partners, a nonprofit, government-supported organisation that manages tourism in the city (Amsterdam's Destination Management Organisation – DMO):

> You see this tipping point where the visitor economy causes more harm for the locals than it adds value, and then you're in an unhealthy situation. We need to change everything we have on offer in the city centre if we want to bring the balance back to living, working and recreating.

Many destinations have suffered from the impacts of excessive mass tourism, known as overtourism (Peeters *et al.*, 2018). The emergent situation drove destinations out of balance. Some destinations try to rethink and reinvent themselves, shifting the perspective towards local communities, adapting approaches in line with responsible or regenerative tourism and focusing on creating value over volume so to maximise societal benefits. Applying scenario planning methodologies can help to envision multiple possible tourism futures and discuss the tourism futures to achieve and those to avoid.

In the case of destination development, the situation presented in Figure 3.2 represent a 'bounce back' situation, where a system is brought back to the stable state it was in before it was impacted on by a shock and/or stress factor. Figure 3.3 represent a possible 'bounce forward' situation, when destinations are brought out of balance, pass the thresholds and re-balance in a fundamentally different situation. This is a different take on resilience. Hence, we should be aware that, next to general understanding of resilience, different types of resilience are distinguished in literature. Davoudi (2012), therefore, distinguished between various types of resilience: engineering resilience (single state); ecological resilience (multiple states); and evolutionary resilience (endless system states, as systems are always in a state of becoming). An evolutionary perspective on resilience fits best with the tourism destinations. For destinations it is very hard, if not impossible, to return to a particular state or situation of the past, being caught up in an endless state of becoming (cf. De Roo & Boelens, 2014). Nevertheless, bouncing forward from one situation to a fundamentally different one is

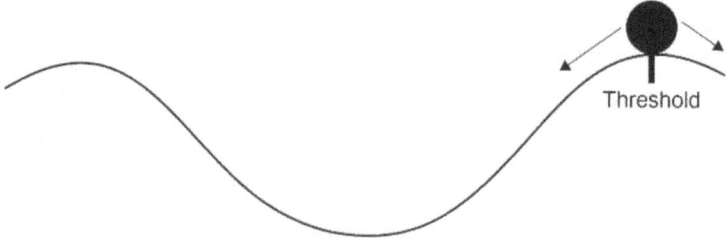

Figure 3.3 Tipping point

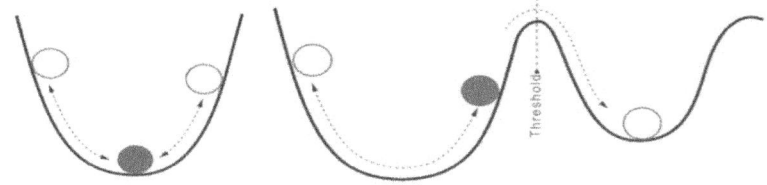

Figure 3.4 Bounce back and bounce forward

a process in itself. In tourism practice, both graphical representations of Figure 3.4 can be found. For many destinations, coping with shocks and stresses is about recovery, about returning to business as usual and about bouncing back to previous system states. This could be desirable by stakeholders because of long-term contracts, vested interests, sunk costs, familiar networks and routines. For other destinations, who aim to build back better after crisis situations, this is not an option or not desirable. The challenge is how shocks and stresses can inspire a transition to a new, better, improved system state that fits better with the ways in which society evolves.

3.4 Resilience by Whom?

The question of *resilience by whom* is an important one but also a complex one. To answer this question, we must further explore complex adaptive systems theories and discuss the concept of adaptation that goes to the core of resilience. Adaptation is the *'iterative and gradual, stepwise process'* that *'might take a series of adjustments over a period of time before systems are fundamentally changed in structure, function, organisation and identity'* (Hartman, 2018b: 157). Alternatively stated, it is the process in which structures, functions and identities of systems are changed in order to deal with pressures that stem from the contextual environment and to move to enhanced forms of organisation and performance (Axelrod & Cohen, 2000; Heylighen, 2001). By means of adaptation, systems are able to change and over time events transform, which makes them resilient. The process of adaptation is driven by individual (change) agents and their actions through which they seek to improve their situation and make (small scale) adjustments to (parts of) a system. Whether they are able to do so and whether they are successful in doing so depends on the individuals' capacities, the system's characteristics (e.g. supportive, restrictive governance regime) and the contextual environment (e.g. forces enabling or constraining adaptation). In the context of tourism destination, who are these agents then? Agents are the entrepreneurs, firms, as well as organisations and public institutions that all seek adaptation – for instance aiming to achieve competitive advantages, improved performances by (re)developing new tourism products and experiences, stimulating sustainability, enhancing liveability, etc.

The collective efforts on these agents influence how systems evolve and whether systems turn out to be resilient. The underlying process of adaptation can be messy. The adaptive behaviour of one agent can be seen as unproductive from the perspective of others. Also, out of the interactions between agents, emergent (organisational and institutional) structures may emerge that support some actions over more productive others. Through constant interaction and (co-)adaptation agents get to learn about their interdependencies and, over time, find combinations that may not be the maximal outcome for each individual but could be the optimal for the collective. Adaptation is, therefore, an endless process doing, achieving, failing, interacting, learning, adjusting and (self)organising. This brings Kauffman (1993: 33) to the observation that *'many parts and processes must become coordinated to achieve some measure of overall success, but conflicting 'design constraints' limit the results achieved'*.

3.5 Concluding Remarks

In Hartman (2018b, 2020, 2021), governance implications of building adaptive capacity for resilient tourism destinations are explored. In these contributions, it is emphasised that adaptive capacity is not a given. Building adaptive capacity, for the sake of resilient destinations, is a complex process that requires much resource investments to ensure key conditions are in place. Here, scenario planning can make a significant contribution. Scenario planning methodologies include a number of aspects that fit well for building adaptive capacity and building resilience.

- First, scenario planning typically includes an understanding of trends and development at multiple levels (Postma & Papp, 2021) which is crucial for enhancing a sense of environmental sensitivity (Hartman, 2018b) and understanding possible forces driving change and their impacts.
- Second, scenario planning brings stakeholder together, to provide impact, share perspective and engage in joint leaning, which is crucial to enhance the connectivity and interaction between stakeholders.
- Third, scenario planning helps to develop perspectives on possible future states of destinations, which is crucial to determine whether, and to what extent, destinations are out of balance, if re-balancing (bounce back) is possible and if tipping points are in sight, that drive or require the fundamental rethinking of the tourism industry and redesign of tourism destinations.
- Fourth, scenario planning helps stakeholders to develop individual, as well as collective, perspectives on future situation to achieve

(best-case scenarios) and futures situations to avoid (worst-case scenarios) and the variants in between. Such insight can motivate agents to take action and, thereby, exhibit adaptive behaviour that is crucial over the resilience of the overall system.

In any case, one should be very aware that there is no one-size-fits-all approach when it comes to applying scenario planning and building resilience. Issues and challenges are often (highly) context-specific, meaning situations can differ greatly from destination to destination. Hence, one should be very explicit about the *resilience of what*, *resilience to what* and *resilience by whom*. In the following chapters we will show and discuss how scenario planning methodologies can be of assistance to building resilience. Methodological steps can be similar from case to case, but sometimes the context-specific situation requires tailor made solution by making adaptations to steps or additional steps as will be evident from Chapters 7 to 12.

3.6 Discussion Questions

(1) In tourism practice, resilience of often (mis)used as a synonym for recovery or resistance to a particular type of (short term) change whereas it is an ongoing process and requires ongoing effort to build and maintain resilience. How is resilience used in tourism practice and how does this relate to resilience as a concept?
(2) Considering the question 'resilience of what', first try to answer the basic questions such as: what is the destination under consideration? What are its boundaries? Who are its stakeholders? Who is (co)responsible?
(3) What are we building resilience against? Factors that cause changes differ per destination. The more explicit one can make what causes (unforeseen, undesirable) change, the better.
(4) Considering 'resilient by whom', to what extent are stakeholders familiar with the concept and understanding of resilience? To what extent is it part of their interest or vocabulary and what does that say about their possible actions toward building resilience?

4 Methodologies of Scenario Planning and Strategic Foresight

Learning Points

- The future can be interpreted as a linear continuation of past and present. This single future is expressed by means of predictive scenarios.
- The future can be interpreted as plural, as the outcome of the complex interplay of uncertain developments. These multiple futures are painted in imaginary explorative scenarios.
- Any type of scenario may act as a source of inspiration for businesses and organisations to become more resilient or future proof.
- If scenarios are linked to strategy development or visioning, this is referred to as strategic foresight.

4.1 The Urge to Visualise the Future

Over more than 3000 years, humankind has been concerned and awed about the future. Over time, they have used different ways to envision the future, as a reflection to prevailing beliefs. Gidley (2017) describes the evolution of futures thinking extensively. From before 1000 BCE to around 500 BCE the future had been painted by prophets, sibyls and shamans. From around 500 BCE to the Renaissance, by philosophers and (medieval) scientists in the form of utopian, mythical, religious and scientific visions. With the birth of modern science and rational thinking during the Enlightenment (18th century) the future was envisioned by means of rational–futuristic fiction and scientific predictions. During the second half of the 18th century, and the beginning of the 19th century, the Industrial, American and French revolutions resulted in dystopian future visions and apocalypse fiction, while during the High Romantic period in the mid-19th century, philosophers focused on creating humanistic ideas of human progress, in the form of cultural and intellectual futures. Since the mid-19th century, theoretical approaches to the future were introduced, such as the evolution theory (Darwin), the idealist communist society (Marx and Engels), social engineering theory (Spencer), theories of social evolution and positivism (Compte). Unbridled optimism about the future, confidence in technology and a belief in endless change led to the birth of a new genre of science fiction that included both utopian and dystopian futures. This spurred the development of social and technological forecasting, which became rather popular until World War I (WWI) in Europe and the US. Futures oriented forms of education were introduced (such as, Montessori, Steiner, Dewey). The launch of the radically new scientific idea of multiplicity of time pressurised the linear time conception. Due to World War I the narrative of progress was put under pressure and the future started to be visualised in dystopian novels that focused on the dangers of technological civilisations, obsession with control, awareness of failure of totalitarian regimes. After the war anxiety about the future broadened and from a wide array of disciplinary perspectives over 100 monographs were written about the future under the title *Today and Tomorrow* (Gidley, 2017).

4.2 The Birth of Scenario Planning (First Era)

After World War I the urge was felt to find ways to better understand, predict and plan for the future. In 1929, the Research Committee on Social Trends in the US was founded, with the aim to study social issues in the US and to chart trends and extrapolate these into the future (Peters & Woolley, n.d.). Thanks to a generally optimist attitude during the reconstruction period after World War II, the idea of the achievable society was thriving, which caused an increase in (long-term) state planning, mainly in relation to military planning. To support them, special bureaus were established. In 1945, the RAND corporation was founded as a leading think tank for military-strategic planning but also for social policy issues (RAND, 2022). RAND envisioned hypothetical futures, referred to as scenarios, as the outcome of a chain of hypothetical events and decisions. By understanding the possible consequences of specific decisions, the scenarios were used to develop strategies and long-term policies. As such, the RAND corporation is considered to be the first to apply scenario planning to governmental policy. Rescher and Kahn at The RAND corporation developed and perfectionated prediction and forecasting methods with the help of mathematics and computer power, which gave the US a dominant position in a predictive approach to forecasting the future and a belief that the future can be domesticated and colonised if predicted well. Later, scientists in the US and the USSR further developed the approach with the help of mathematics, modelling, simulation, gaming, etc., to make the Cold War more predictable and manageable.

The approach, introduced and developed by the RAND corporation, assumes a single, surprise-free and achievable future that can be predicted by means of robust empirical scientific methods. It is based on a positivist/empiricist epistemology and reductionist thinking. Just like in physics, it is believed that the world is arranged by universal truths and objective regularities from which human behaviour can be predicted. Within this approach scenarios are developed based on an analysis of the past (historical data, trends analysis) and an extrapolation of these data into the future, in which uncertainty is reduced to statistical uncertainty. This linear mode of thinking results in predictive or extrapolative scenarios that describe what is likely to happen ('what *will* happen'). In scenario planning, these predictive scenarios are often used as the baseline, or the baseline scenario. It is common to develop these types of scenarios with a small team of experts and scientists, while the relationship between the researchers and the subject under consideration is distant. The approach is referred to as 'speaking truth to the power' and emphasises the product: technically sound forecasts. Major advantages of the approach are perceived objectivity and value neutrality. Disadvantages are a narrow focus and lack of contextual awareness and the fact that trends are sees as unavoidable.

Predictive or explanatory scenarios take the form of a forecast, based on a set of assumptions concerning variables with a stable or recurring pattern that are used to predict a linear development of the issue under investigation. Future uncertainty is interpreted in statistical terms, as a statistical bandwidth around the central projection (Table 4.1).

Table 4.1 Comparison between predictive and explorative scenarios

Predictive scenarios	Explorative scenarios
The future is single and certain – what *will* happen	The future is plural and uncertain – what *could* happen
Past explains the present	The future is the *raison d'être* of the present
Based on linear thinking	Based on discursive thinking
Study of historical data	Study of the future (early indications of change)
Identification of trends (trend analysis)	Identification of patterns, processes and drivers of change (causal layered analysis)
Quantitative, objective	Qualitative, subjective
Statistical and stable relations	Dynamic and emerging structures
Uncertainty is interpreted as statistical inaccuracy of the forecast	Uncertainty is starting point for the exploration of multiple futures
Output is (single) extrapolative scenario	Output is (multiple) explorative scenarios
Processed by experts	Processed by a large group of participants
Passive or adaptive attitude (the future will be)	Active and creative attitude (the future is created)
Product/output is key (the forecast)	Collaborative process is key (strategic learning)

Source: Adapted from Lindgren and Bandhold (2009).

4.3 Second Era of Scenario Planning

During the 1950s and 1960s, the closed system, mechanical worldview was increasingly under attack, due to new insights into how systems operate. Systems were perceived to be complex and self-adaptive, with interdependencies between its elements, a high level of dynamics and unexpected behaviour and chaos because of direct and indirect feedback. These insights, coupled with the emergence of post-positivist science based on pluralism and the belief of multiple subjective realities as a result of the mental constructs of individuals, and so of multiple futures (Kuhn, Popper, Habermas) gave rise to a counter-movement to prediction and forecasting in Europe and elsewhere that gradually put the prevailing idea of the achievable society under pressure during the 1960s and 1970s. According to this new approach, the complexity of human affairs and a human's bounded rationality hamper the ability to trace the future and to predict it accurately and in detail. The new approach, addressing unpredictability and multiplicity of the future, was established by academics from a wide variety of disciplines, such as systems science, sociology, journalism, theology, media and peace research, and initiated a diverse range of qualitative and interpretative methods better suited to social science research. The publication of the report *Limits to Growth* in 1972 (Meadows *et al.*, 1972) is often regarded as a booster of the new paradigm. The report used computer simulations and model calculations to explore the long-term effects of population growth and industrialisation. In response to the new developments in futures thinking, the World Futures Studies Federation (WFSF) was established in 1973. Various national bodies were established to study the future by means of scenarios. This approach fits with broader developments of thinking about democratisation, knowledge development, learning and decision making.

This approach takes the complexity and future uncertainty of tourism and its context as a starting point. In fact, the system is perceived to be so complex, dynamic and ambiguous that it cannot be known in its entirety, let alone how it will evolve in the future. Due to uncertainties, the system may develop in multiple ways and so the future is perceived to be plural. It cannot be predicted, yet only be studied (futures studies, futurology or futurism), explored and imagined. The best way to explore the future of the 'infinite' system is to create a few windows through which we can take a closer look at it and try to understand it better. These windows are referred to as explorative scenarios. Each of these scenarios paints a different picture of the future. Explorative scenarios describe what *could* happen. Due to the complexity and ambiguity of the system, the development of this type of scenarios is based on holistic thinking and a constructivist epistemology. Ideally, people that represent a range of expertise, disciplines and stakes are collaboratively developing the scenarios. The approach is process oriented. Participants share knowledge, insights, opinions and experience, challenge each other's paradigms and 'dominant thinking' and develop consensus on how the system could be understood. The process shows a close interaction between theory and practice and direct involvement with social practice. Besides, there is a close relationship between the participants and the subject under consideration. The approach is referred to as the 'arena approach'. The development of explorative scenarios is based in non-linear or discursive thinking, with the aim to develop comprehensive understanding of the system context, that is closer to reality and more focused on application. In contrast to the singular predictive scenario, the explorative scenarios are sometimes referred to as alternative scenarios that complement the baseline.

Explorative scenarios are paintings of the future in words and probably also in images, to trigger both halves of the human brain. They assume the intervention of several key events or conditions which will have taken place between the time of the original situation and the time in which the scenario is set. To make the scenarios plausible, ideally the chain of these events is described in a logical way. This type of scenario can be related to different time horizons but are most powerful if the time horizon is chosen to be long, and so uncertainties are large (Table 4.1).

To make scenarios more powerful, they can be illustrated with photographs, cartoons, video clips, films, dedicated booklets, performances, scenario rooms or virtually in 3D. Serious gaming may be used to simulate long-term effects of policy measures or strategies. This allows participants to engage with the scenarios in an interactive way.

4.4 Third Era of Scenario Planning – Consolidation of Scenario Planning

The growing interest in the pluralistic way of scenario planning was fuelled by the World Futures Studies Federation (WFSF) and the United Nations Educational, Scientific and Cultural Organisation (UNESCO) and resulted in an increasing number of conferences, courses, workshops and academic journals, and so a wide dissemination and exchange of ideas, concepts and methods.[1] This resulted in a democratisation process of scenario planning and a growth in the number of futurists across the globe.

Scenarios (baseline and alternative scenarios) provide information about how the future may evolve (predictive scenario) or might evolve (explorative scenarios). Such scenarios offer an interesting glimpse into the future, but for business or organisations they only become functional when they use the scenarios to prepare itself for that future using targeted strategic actions. With this we enter the field of strategic foresight. Strategic foresight is about strategy development, decision making and preparing for action.

In the 1970s the idea of scenario planning was transferred to the business world when Wack introduced it to Shell. Thus, scenario planning became more closely connected to strategy. Later the approach was popularised further by De Geus (1988), Schwartz (1991) and Van der Heijden (1996, 2004), and scenario planning started to spread within the business world. The oil crises during the 1970s made the scenario approach less popular as it was considered to oversimplify thinks. However, in the 1980s, the interest in scenarios for planning purposes increased again. It had the wind in its sails when Michael Porter (1980, 1985, 1990) started to emphasise the importance of the external business environment in strategic planning, which was uncommon those days. He asserted that strategies should be regarded as dynamic rather than static. Besides, scenario thinking had been increasingly promoted as a means to improving decision making with regard to environmental uncertainties. During the 1990s turbulence in society caused another boost of scenario thinking and using scenarios for strategic planning. Several consultancy firms developed their own style and sometimes even their own terminology.

Recently, the emergence of a fourth era in the development of scenario planning can be identified. This fourth phase demarcates a shift away from perceiving scenario planning as a practice driven methodology, shaped by practitioners for practitioners, which has resulted in a broad variety of methods and related jargon. During this era the attention has increasingly shifted to the theoretical underpinning of scenario planning, and so to scenario planning as a theory builder rather than a methodology. This view of scenario planning as a theory builder was extensively discussed in Chapter 2 of this book.

4.5 Strategic Foresight

The term strategic foresight was introduced in 1996 by Martin (1996, 2010) as the most appropriate translation of the French term 'prospective'. The idea of 'prospective' was introduced as early as 1957 by Berger (Berger, 1957; De Jouvenel, 2018). If a business or organisation does not want to be driven to unpredictable directions, but wants to be resilient, adaptive, future proof or even serve as a driver of change, then strategic foresight is needed. Strategic foresight, like 'prospective', acknowledges that there are multiple futures possible, and it puts an emphasis on group processes, participatory debate and on the results of that process in terms of action. It is believed that developed, possible and desirable futures can only be of interest to an organisation or business if they are based on intimate knowledge of the dynamics of the environment and so destined to have a real influence on action.

It is often more comfortable projecting from the past rather than imagining the future and strategic planning approaches often incorrectly assume that the future will be an incremental iteration of the past or the present. However, strategic foresight assumes that humans have the will and capacity to influence the future in order to favour the desirable, and the capacity to reflect upon multiple possible futures, reflect on the implicit paradigms underlying their decisions and question and possibly modify their thinking. That the future is not conserved as inevitable and predestined but as something that is created by key actors and those prepared to sacrifice in order to make their plans succeed.

4.6 The Use of Strategic Foresight

'Strategic foresight is an organised, systematic way of looking beyond the expected to engage with uncertainty and complexity' (UNPAN, 2021). According to the OECD (2019), strategic foresight equips organisations and businesses with the capacity to stay ahead of time, by continually exploring and preparing for the future in order to navigate, adapt and shape it by means of better policies and strategies. According to Conway, strategic foresight is the ability to take a forward view. It is a competence to think systematically about the future to inform decision making today. It enables action to be taken today with reference to, and within the context of, the future. It is the capacity that emerges in organisations when people work together to build strategic processes that take a long view (Conway, 2007).

Basically, according to Ozbekhan (1977, in Godet & Durance, 2011) there are four stages in attitudes toward the future, such as painted in scenarios:

- A passive attitude, that is, accepting change without challenging it.
- A reactive attitude, that is, waiting for the alarm bell to ring and only then taking action.
- A pre-active attitude, that is, anticipating changes in the future environment so as to take advantage of such changes and to prepare the organisation or business.
- A pro-active attitude, that is, designing a desired future and then creating desirable changes through action.

Strategic foresight requires a pre-active and pro-active attitude, as it can help organisations and businesses to innovate. However, in times of crisis, such as during the COVID-19 pandemic, a reactive attitude prevails.

4.7 Strategic Foresight Process and its Components

According to the European Foresight Platform, strategic foresight is an open, participatory and action-oriented process in which the participants representing diverse voices and perspectives, collaboratively think about the future (future intelligence), debate the future (strategy development) and shape the future (turn strategies into action). These components of the strategic foresight process are in line with the Greek triangle of prospective, as conceived in 1995 (Godet & Durance, 2011): thinking the future refers to anticipation, debating the future to appropriation and shaping the future to action (see Figure 4.1). A business or organisation with a lack of anticipation, ongoingly needs to respond to pressing problems, draw resources away from considering its long-term development and will not be able to get the better of the external circumstances.

The process to arrive at strategic ambitions and strategic courses of action is much more important than its outcome because the business environment is subject to constant change, preparation will never be finished and it will need to be adapted regularly. To become more resilient or future proof to the continuous changes in the environment, an ongoing strategic learning process is required, in which the stakeholders interact with each other. Since there are no statistics about the future, personal knowledge, experience and judgment are the best possible alternative. That is why it is important to gather as many opinions as possible. As long as someone can bring in a fresh perspective, they are welcome. During this process, knowledge, opinions, interests and experiences are shared, exchanged and connected to eventually reach

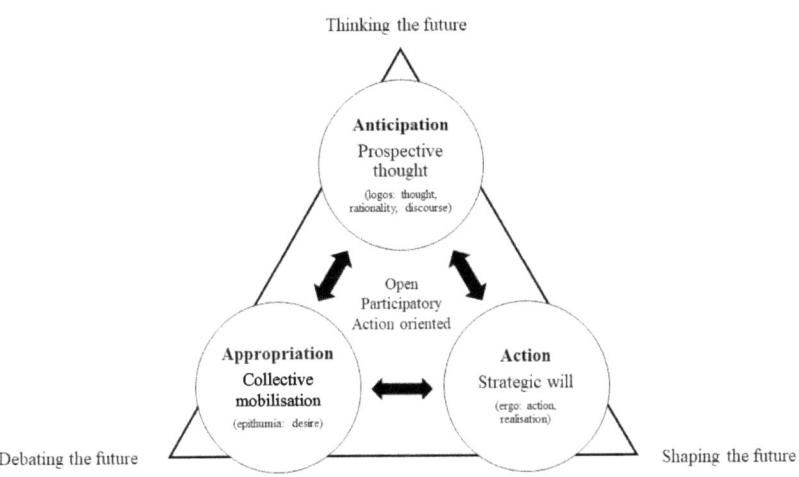

Figure 4.1 Components of strategic foresight
Source: Adjusted from European Foresight Platform (n.d.) and Godet and Durance (2011).

a consensus. To stretch the thinking of everyone involved, there is a need to challenge each other's paradigms and borders of thinking, broaden the understanding of the context and improve the competencies, skills and adaptive power to deal with uncertainties. The process requires imagination, empathy, creativity and outside-the-box thinking.

4.8 Scenarios as a Key Tool in Strategic Foresight

Scenarios act as a key tool in strategic foresight. Strategic foresight starts by exploring the future by means of scenarios (anticipation), typically beyond the planning horizon, before proposing various strategic orientations (appropriation) and putting them into action. Explorative scenarios disrupt dominant thinking and reveal new perspectives with which businesses and organisations are able to:

- anticipate better: it can assist at an earlier stage with identifying of, and preparing for, opportunities and challenges emerging in the future;
- innovate policy: it can spur new thinking about the best policies or strategies to address these opportunities and challenges; and
- future proof: it can help to stress-test existing or proposed strategies against a range of future scenarios.

Lindgren and Bandhold (2009), assert that scenarios can be used in four ways:

- for business development/concept development (focus on new business and purpose of action);
- for strategy development/organisational development (focus on new business and purpose of prerequisite for change);
- for new thinking/paradigm shift (focus on old business and purpose of action); and
- for risk-consciousness/need for renewal (focus on old business and purpose of prerequisite for change).

Since strategic foresight is an ongoing cyclic learning process, the development of scenarios is not a one off. Due to ongoing changes in the environment, it is important to monitor/scan the horizon permanently, to be cautious and alert to early indications of new (emerging) developments. This could trigger decision makers to adjust the scenarios, if necessary.

4.9 Outcome of Strategic Foresight

During the foresight process, first, a normative goal-oriented scenario may be created. This type of scenario is also referred to as an aspirational scenario. Inspired by the explorative scenarios they describe the collective desire of the organisation or business, or they synthesise key ideas of the explorative scenarios. Thus, they describe a desirable future, and their purpose is to show how certain objectives can be realised, as well as the paths to achieve these objectives.

Eventually, strategic foresight results in strategic courses of action that can be put in practice. This could be a new vision, concept, business model, product, service, etc. Ideas could be directly deduced from the explorative scenario's or indirectly via a goal-oriented or aspirational scenario. Strategic courses of action could be risky, semi-robust or robust. Risky or betting strategies refer to strategies that fit with only one of the scenarios – it is possible to have a betting strategy in place for each possible scenario and to upscale or downscale such strategies, if early indicators of the scenarios require so. Robust strategies are strategies that withstand the organisation or business for any of the scenarios, so for any future, while semi-robust strategies prepare it for two or three scenarios. Due to the confidentiality and responsibility surrounding strategic choices, the number of people involved in defining strategic courses of action may be limited. This concerns, for example, the management or board of directors of an organisation or company.

4.10 Concluding Remarks

Interestingly, all our knowledge is about the past, while our decisions are related to the future. However, an environment that is volatile, uncertain, complex and ambiguous (VUCA), urge businesses and organisations to learn about the future and to anticipate emerging opportunities and threats. Scenarios and strategic foresight provide a useful approach to do so.

Strategic foresight goes beyond linear problem-solving approaches and a focus on single issues. Instead, it is based on a collaborative process that cuts across the traditional boundaries of strategic or policy domains within organisations, governments or businesses, and incorporates societal, economic and environmental perspectives.

Strategic foresight is an approach, a methodology, that is most effective when there is support of the organisation and of its internal and external stakeholders. This implies that the ideal moment to perform scenario planning is at the start of a new policy cycle, or when major changes are anticipated within the organisation or its environment that ask for a response.

A study by Hamel and Pralahad (2005, in Godet & Durance, 2011) shows the relevance of the strategic use of scenarios. They compare more and less successful businesses with each other. Successful businesses spend less time worrying about how to position themselves in the existing competitive space but more time creating fundamentally new competitive space. They are capable of imagining products, services or entire industries that did not yet exist and then to create them. The laggards, on the contrary, are more interested in protecting the past than in creating the future.

Dedicated application of scenario planning and strategic foresight to the domain of tourism is rather new. It started back in 1993 with Ian Yeoman's work at VisitScotland, the national tourism organisation for Scotland. VisitScotland appointed Ian Yeoman as full-time professional/futurologist and scenario planner at its Futures Department to undertake futures thinking and scenario planning. Over the years, VisitScotland developed a scenario planning process based on three key ingredients: a scenario planning group to construct, develop and implement scenarios for VisitScotland and its partners, environmental scanning as a qualitative process to capture shocks, surprises, trends and drivers that will influence and shape tourism in a systematic and sensible manner and the Moffat Model, to quantify scenarios (Yeoman & McMahon-Beattie, 2005).

VisitScotland's work inspired the tourism academy at the predecessor of NHL Stenden University in the Netherlands to establish a European Top Institute specialising in the scenario planning to support the European tourism industry. Since the European Tourism Futures Institute (ETFI) started its operations in the autumn of 2010, Yeoman has been acting as visiting professor and as supporter and mentor for its staff. Based on the principles of scenario planning and strategic foresight, as described in this chapter and an expanding portfolio of projects, the ETFI has developed and fine-tuned its own approach since 2010 (Postma, 2013a, 2015b). This approach will be outlined in the next chapter.

Since Ian Yeoman was appointed by the University of Wellington, New Zealand, as associate professor and academic researcher in the domain of tourism futures, he has developed into a world-leading academic, practitioner and speaker on scenario planning in tourism and hospitality. (Yeoman, 2012a). Since 2023 he works as professor of Innovation, New Phenomena and Disruptions at the Hotel Management School Leeuwarden at NHL Stenden University.

4.11 Discussion Questions

Consider a specific issue or uncertainty that puzzles you, 'that keeps you awake at night'.

(1) Shortlist a few major trends that affect the future of this issue.
(2) How would the issue look in 10 years' time from now, if these trends would continue in the same way?
(3) Which unexpected events, innovations, etc. could disturb this development? So, what could happen to the issue instead?

Note

(1) It should not be overlooked that, after the oil crises during the 1970s, businesses turned back to the since the 1990s, after the end of the Cold War, the emergence of a neoliberal view led to a renewed emphasis on the predictable single future paradigm of the USA.

5 The Application of Scenario Planning in Tourism

Learning Points

- Using scenario planning with policy and politics.
- Using scenario planning for crisis management.
- Using scenario planning with trends analysis and product development.
- Using scenario planning and creativity to design future hotels.
- Using scenario planning to explore the future of work and technology.

5.1 Introduction

In the present era, characterised by uncertainty, innovation and crisis, increasingly, emphasis is placed on the use of scenario planning techniques because of its usefulness in times of uncertainty and complexity (Amer *et al.*, 2013; Cordova-Pozo & Rouwette, 2023). This was demonstrated by the use of scenario planning as a methodology to make sense of COVID-19 scenarios and recovery with a focus on regenerative tourism. For example, in New Zealand, the government set up a Tourism Futures Task Forces in order to create a strategy for the tourism post-COVID-19, and further projects included addressing the issue of future workforces and climate change (Yeoman *et al.*, 2022). In the Pacific, the Asian Development Bank commissioned a number of projects regarding sustainable tourism as a driver of change away from resort-based tourism, which illustrated the impact and change through scenarios (ADB, 2021; Becken & Loehr, 2022). The European Tourism Futures Institute (ETFI) constructed a scenario set for the visitor economy in the Netherlands (see Chapter 7) (Postma *et al.*, 2020a). Clark *et al.* (2022) reported the use of scenario planning as a tool to drive post-recovery scenarios in Arizona. Seyitoğlu and Costa (2022) noted the use of scenario planning in many destination management organisations, as a tool for recovery post COVID-19 in Europe.

One of the roles of government is preparing for the future, through planning for it. Strategic foresight is a central function of government planning, whether it is understanding demography changes, planning for healthcare provision, how would rising sea levels impact on spatial planning or understanding external political threats that would shape defence spending (Boston, 2017). It is the same for tourism, in order to create a desirable future for tourism in any destination, it is important to understand the external forces that will shape demand and supply (Dredge & Jenkins, 2007).

Long-term policies can be framed regardless of the political philosophies but are often shaped by the political outlook of the governing party. A government is not going to advocate the direction for tourism if it goes against the value system and belief system of that party (Albrecht, 2017). Whereas on the other hand, for instance, can democratic institutions, policymaking processes and analytical frameworks be designed in ways that increase the likelihood of long-term interests receiving adequate attention which go beyond political philosophies? How can the chances of short-sighted policy decisions – ones that threaten or undermine tourism's long–term wellbeing – be minimised? This is often achieved when politics is shaped by coalition governments or a pragmatic approach to policy (Boston, 2017).

Another reason why foresight tools, such as scenario planning, are used is to overcome the tendency for governments to give excessive weight to short-term interests and considerations, thereby putting future interests at risk (Boston, 2017; Dredge & Jamal, 2015; Dredge & Jenkins, 2011). This tendency is variously referred to as short termism, political myopia, policy short sightedness, a present bias and a presentist bias. The widespread nature of this phenomenon suggests that it constitutes a general *disorder*. In other words, it represents a systemic governance problem, not simply a policy problem or a problem specific to a particular policy domain. As such, it cannot be ameliorated merely by changes to discrete policy settings. It requires a more holistic and comprehensive response.

5.2 Case Study One: The Policy and Politics of New Zealand Tourism

5.2.1 Overview

Tourism is cross-sectoral and there are many stakeholders with responsibilities for tourism strategy and management. Nevertheless, right at the heart of tourism policy is politics and political theory (Albrecht, 2017), therefore, tourism policy represents the political philosophy of the governing party. In New Zealand, the government is dominated by two political parties, The National Party, which advocates capitalism and business, and the Labour Party, which advocates socialism, community and a fairer distribution of wealth. As the country has a proportional representation for elections, political parties never govern by themselves, as they are dependent on minority parties, that is, New Zealand First, ACT, Green Party or Māori Party. So political philosophies become compromised or diluted, therefore, politics often takes a pragmatic and pluralistic philosophy (Boston, 2017; Sheptycki, 2020). In addition, policy is engrained in the Treaty of Waitangi, based upon the indigenous rights of Māori (Matunga et al., 2020; Mika & Scheyvens, 2022; Ransfield & Reichenberger, 2021; Scheyvens et al., 2021).

In general terms (excluding the COVID-19 period), tourism in New Zealand represents nearly 1 in 10 jobs, 9.1% of the country's GDP and 18% of foreign exchange earnings. The country's tourism brand, *100% Pure New Zealand*, is world leading. However, visitor demand and global tourism trends are changing. The *Tourism 2050* project, commissioned by the Ministry of Tourism and the Foundation for Research in Science and Technology, set out to envision the future of tourism in New Zealand by asking the question, *'What will New Zealand tourism look in the year 2050?'* The research project produced four scenarios, *Manaakitanga, An Eco Paradise, Perfect Storm* and *The State of China* (see Figure 5.1), which all follow different pathways and are constructed upon realistic circumstances. The purpose of the scenarios was to provide a foundation to encourage and promote discussion, in order to understand New Zealand's tourism future and to 'think differently'. The scenarios provided a comprehensive analysis of the external and internal challenges New Zealand's tourism industry would face in the future and conclude with a series of strategic recommendations which are the basis of a tourism national plan. The scenarios offer a detailed account of that scenario planning process, content and outcomes, thus, providing policymakers with a blueprint of how to do scenario planning in the tourism context and serves as a guide to the future of tourism in New Zealand.

5.2.2 Political philosophy and scenarios

The *Eco Paradise* scenario was based on collectivism drawing on the principles[1] of the New Zealand Labour Party as:

- the natural resources of New Zealand belong to all the people and these resources, and in particular non-renewable resources, should be managed for the benefit of all, including future generations;
- all people should have equal access to all social, economic, cultural, political and legal spheres, regardless of wealth or social position, and a continuing participation in the democratic process; and

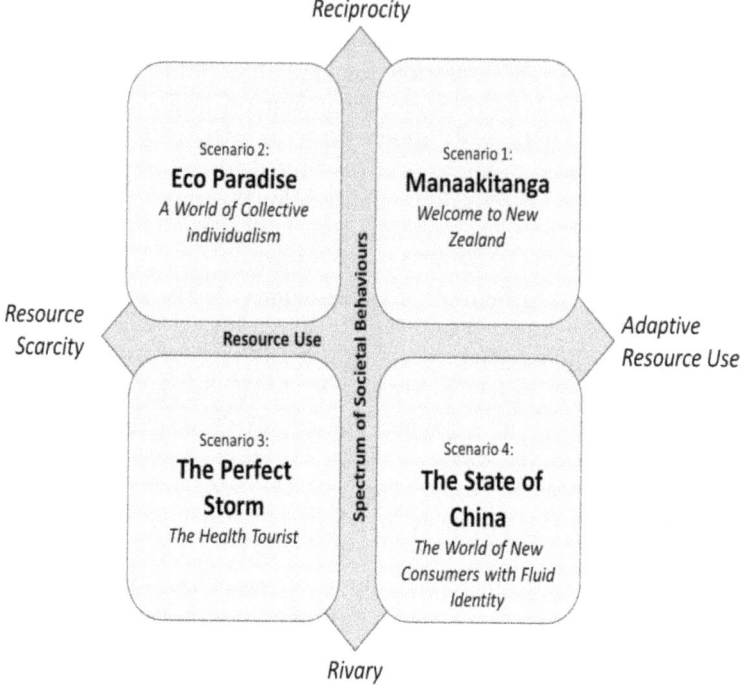

Figure 5.1 Scenario planning matrix – New Zealand

- cooperation, rather than competition, should be the main governing factor in economic relations, in order that a greater amount and a just distribution of wealth can be ensured.

Whereas the scenario *The State of China* scenario advocated values of the New Zealand National Party[2] as the individualism and capitalism, which was the basis of the as:

- individual freedom and choice;
- personal responsibility; and
- competitive enterprise and rewards for achievement.

Manaakitanga was designed to represent the principles enshrined in the Treaty of Waitangi and aligned with the policy of Māori Party.[3] The scenario took a strong utopia, regenerative and community tourism perspective (Yeoman et al., 2015b). Whereas, *Perfect Storm* was about economic events and near collapse which was influenced by a series of events, such as the global financial crisis and SARS (Page et al., 2006; Pine & McKercher, 2004).

5.3 Case Study Two: Preparing for a Crisis

Surprising events – or what Nassim Taleb (1997) calls 'black swans' – are inevitable. Such events lie beyond the realm of normal expectations. To complicate matters, governments face not only the 'known unknowns' – such as when the next major earthquake, drought, flood or financial crisis might (Page *et al.*, 2006; Yeoman & McMahon-Beattie, 2005) – but also 'the ones we don't know we don't know', as Donald Rumsfeld, a former American Secretary of Defense, famously put it (Pawson *et al.*, 2011). As Yeoman *et al.* (2005) points out, Foot and Mouth Disease, the Gulf War and 9/11 were 'known unknowns' rather than 'unknown unknowns' as these external events are well documented in the science literature, and government departments prepare for them (Gershwin & Crowley-Cyr, 2021). The unknown is that tourism at government level has historically not engaged in preparing for the unknowns using foresight techniques, such as scenario planning or environmental scanning. Historically, destination leaders were focused on a rail-track approach to the future, based upon the past and exponential growth. The reason for this is that government expenditure on tourism is fundamentally on marketing rather than on governance and leadership (Hay, 2019, 2021). COVID-19 and tourism unpreparedness are classic examples of this (Bhaskara & Filimonau, 2021).

5.3.1 How VisitScotland prepared for war?

5.3.1.1 Overview

VisitScotland, the national tourism agency for Scotland, used a scenario-planning process to untangle the complexity of the forthcoming war in Iraq. The scenarios explored the impact of such a war on tourism against a backdrop of an economic environment of failing equity markets and GDP. Scotland was on the verge of a recession, and VisitScotland wanted to know how war would affect this economic environment and, simultaneously, how this would affect different tourism markets. VisitScotland constructed four scenarios: how the *West was won, global Northern Ireland, new dawn* and *into the valley of death*. The scenarios helped the organisation develop policies and actions to deal with contingencies in each scenario. More importantly, the article shows how VisitScotland managed the process, what it did and the policy implications for the future.

5.3.1.2 From ridicule to destination leadership

At the time of the Iraq War, VisitScotland's Scenario Planner, Dr Ian Yeoman, set about constructing a set of scenarios, working with Miriam Galt of BeeSuccessful Consulting (https://www.beesuccessful.com/) to establish what the war would look like and the short- to medium-term impacts on Scottish tourism. The first draft of the scenarios was in fact rejected by the Chief Executive, Philip Riddle, as too extreme and inappropriate.

Ironically, at about the same time, British Army soldiers were stationed at Heathrow Airport and panic set in at VisitScotland regarding the implications of war. Indeed, this was a tipping point that Yeoman had predicted in the scenario Global Northern Ireland (Yeoman et al., 2005a). Malcolm Roughead, Director of Marketing at the time, immediately sensed this and asked Yeoman to draw up a second draft of the scenarios, modelling them from an economic perspective (with the help of www.oxfordeconomics.com) and embedding the scenarios into the organisation.

One of the roles of VisitScotland was to show destination leadership and through the scenarios this was achieved. Yeoman had demonstrated through the scenarios how effective they could be when facing a potential crisis and the 'scenario planning' role was now accepted within the organisation.

5.3.1.3 Communicating in a crisis

Once war became imminent, VisitScotland activated their crisis management and communications strategy to deal with the forthcoming war. At this stage, it was important for the Scottish tourism industry to 'buy in' to VisitScotland's scenarios. The importance of the scenarios allowed stakeholders to 'make sense' of the pending war, which was perceived to be a wicked problem (Huff & Jenkins, 2002). The scenarios allowed stakeholders to be 'comforted', and a 'sense of panic' was prevented.

At the centre of this strategy was a joint action group (JAG) made up of representatives of Visits land, the local Area Tourist Board Network, the Scottish Tourism Forum, British Airways, the British Airports Authority, Scottish Enterprise, the Scottish Executive, the National Trust of Scotland, the Edinburgh Principal Hoteliers Association, the Scottish Retail Consortium and other hotel representatives. The purpose of this committee was to confirm and support actions, agree to key messages and use them in all communications, and coordinate Scottish tourism's response to the war in such a way that there was only one principal voice for the industry. The committee undertook a number of activities, including the dissemination of information through their networks and a dedicated website. They also began a monitoring process that examined changes in booking patterns, tourist behaviour and business and consumer confidence. The communication team at VisitScotland took responsibility for coordinating the activity as actioned by JAG. This committee acted as a unified voice for the industry that was able to bring about considered responses before, during and after the war. Considered responses were extremely useful in communicating accurate and effective information to industry, rather than many of the doom-and-gloom messages that appeared in the press.

5.4 Case Study Three: Trends Analysis and Product Development: The Importance of Family Tourism During COVID-19

Saritas and Smith (2011: 274–275) note that drivers of change are 'those forces, factors and uncertainties that are accessible by stakeholders and create or drive change within one's business or institutional environment'. The use of a 'drivers of change approach' is well documented in futures and scenario planning literature (Luhmann, 1976; McDonald, 2011; van 't Klooster & van Asselt, 2006; van Zon, 1992). Drivers of change or mega tends are overarching forces which shapes the future of society, whereas micro trends are sub trends of these driving forces. (Postma & Papp, 2021) The term 'micro trends' was coined by Mark Penn (2007) in *Micro Trends the Small Forces Behind Tomorrow's Big Changes* and more recently in *Microtrends Squared: The New Small Forces Driving Today's Big Disruptions* (Penn & Fineman, 2018). Micro trends illustrate how the mega trend comes to life within a tourism context. In this case study we illustrate how a trends analysis approach was used by Yeoman, *et al*. (2022b) to understand how COVID-19 altered those trends and subsequently highlighted the opportunity of family tourism during the pandemic (Yeoman *et al*., 2022a).

5.4.1 COVID-19

An important feature of scenario planning is understanding the drivers of change and trends that are shaping the future. COVID-19 was considered a pandemic and was considered a 'once in a century' public health event, which had a profound impact on global tourism. For one country, like many, tourism in New Zealand is the country's largest export earner and with the closure of the country's borders, international arrivals collapsed. Consequently, the focus became domestic tourism and how tourists would change because of COVID-19 at different alert levels.

5.4.2 Trends analysis

Using an evaluation matrix (see Figure 5.2) developed by Flatters and Willmott (2009), the authors set out in May 2020 to show what the trends were expected to be: the slowed trends; the trends that will be arrested and the ones that will advance. The matrix allows destination planners and tourism businesses to evaluate what trends to focus on in a COVID-19 scenario. Fifteen trends were identified by Yeoman *et al*. (2022) which are consumer behaviour orientated. These trends are

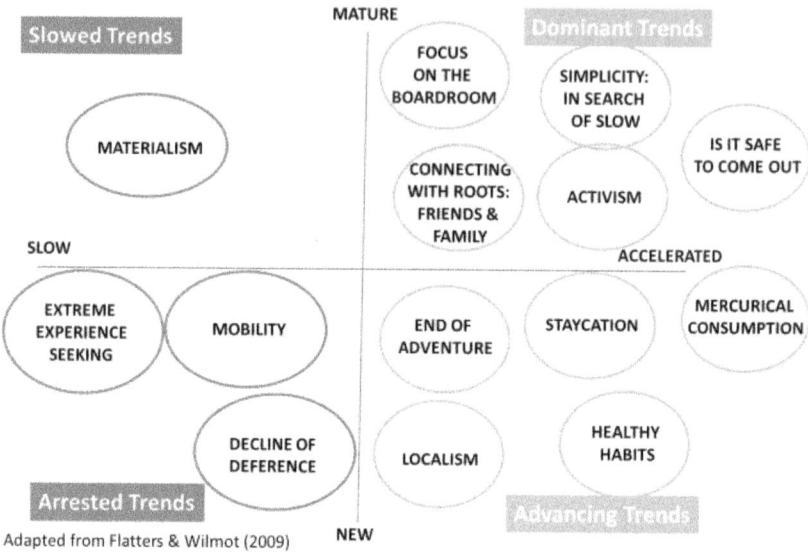

Figure 5.2 Trends matrix

classified as dominant, slowed, advanced or arrested. The trends that are *dominant* are: Focus on the Boardroom; Simplicity in the Search of Slow; Connecting with Roots: Friends and Family; Activism and Safety. The *slowed* trend is Materialism. Trends that are *advancing* include End of Adventure; Staycation; Mercurial Consumption; Localism; Healthy Habits and Presence Free Living. *Arrested* trends include Extreme Experience Seeking; Mobility and Decline of Deference.

5.4.3 Dominant and advancing trends

In the paper *Tourism Trends in a COVID-19 World: A New Zealand Perspective* (Yeoman et al., 2022), a number of trends were identified as advancing and dominant, between May 2020 and 2021. These trends included the following:

5.4.3.1 Trend 1: Simplicity: In search of slow

During an economic slowdown, tourists tend to travel less, stay near home and seek simplicity, such as value-based holidays focusing on basic facilities, meeting locals, lots of free time and bargains (Yeoman, 2016). Here, tourism is about the beach or mountain views.

5.4.3.2 Trend 2: Mercurial consumption

During recessions, impulse purchases dramatically decline. Tourists seek out bargains. This means planned purchases come to the forefront.

Women, or those with responsibility for household budgets, search more deeply for information, wanting to know about everything including activities, experiences, reviews and prices (Factory, 2018c).

5.4.3.3 Trend 3: Localism

Across the globe, the trend of localism has been advancing as tourists (Towner & Lemarié, 2020) and consumers are invited to show their support for all that is local by buying products with proximate provenance, choosing companies which engage actively with local communities or holidaying in one's own country.

5.4.3.4 Trend 4: Staycation

Unsurprisingly, in a recession tourists focus on domestic, rather than international, holidays (Hall *et al.*, 2021) which is known as staycation (Cvelbar & Ogorevc, 2020).

5.4.3.5 Trend 5: Healthy habits

The concept of health has developed from something associated with diet and physical exercise to a holistic entity (Quorin *et al.*, 2020), but is also associated with mental wellbeing and mindfulness (Factory, 2018a, 2018b).

5.4.3.6 Trend 6: Is it safe to come out?

Safety has become of paramount importance to tourists in COVID-19 scenarios (Singh, 2020) based upon an unprecedented level of public fear, which Zheng *et al.* (2021) call travel fear. Fear is a primitive emotional response to a threat, that is, COVID-19 which is incalculable and unpredictable.

Combined, these six trends have had a significant impact on shaping one important segment: *Family Tourism*.

5.4.4 Family tourism, domestic tourism and COVID-19

The primary form of tourism for families is domestic, especially in New Zealand (Trend 4). There are many affordable places for families to go to and have a fun holiday, for example, the Department of Conservation (DoC) campgrounds and holiday parks (Trends 1, 2, 3 and 4). New Zealanders see the annual summer holiday by the beach or lake as a well-loved Kiwi icon and symbol of Kiwi identity (Yeoman *et al.*, 2022a). Additionally, pets are increasingly forming an important part of families and need to be accommodated to travel.

There are changing family structures, with often smaller core families and people living longer, leading to more grandparents enjoying multi-generation holidays and more time with their grandchildren, or grandtravel (Gram *et al.*, 2019). The continuing change in gender roles and parenting behaviour can also have significant influences on family holiday experiences, with, for example, fathers more actively involved with their children and mothers seeking more relaxing 'me time' time away from the family (Schänzel & Smith, 2011) (Trend 1). However, many families have also been priced out of the opportunity to show their children more of the country, for example, most children in Auckland have never been to the South Island but might have visited Australia because this is perceived as cheaper (Trend 2). Families are considered more price-sensitive and budget-minded, but this is not true for all of them. There are also those that usually go on an overseas trip in winter to seek sun but might, because of travel restrictions, consider a snow holiday, providing opportunities for domestic winter holidays (Trend 2).

5.5 Case Study Four: Using Scenario Planning and Creativity to Design Future Hotels

Lub *et al.* (2016: 249):

> explores the use of design thinking as a method to develop scenarios for the future of hotels. Using a Dutch case study, the authors show how a new concept for hotels – the Lifestyle Hub – was created using design thinking as the methodology. The Lifestyle Hub concept provides ingredients to hotel owners, as well as public policymakers, to help understand how future guests may expect to make use of individually tailored, hospitable facilities in destinations around the world. Moreover, design thinking allows researchers and businesses to generate highly differentiated customer-centred, experience-based, business concepts, thus, adding to the toolkit of futures researchers. We conclude that design thinking provides new insights for hospitality and tourism and presents a valuable alternative to current future scenarios approached.

5.5.1 Design based thinking and scenario planning

Right at the heart of the article is a workshop-based method called design-based thinking, which is used to enhance the scenario planning process through creativity, given the rationality and sometimes linear process of scenario planning (Wilkinson *et al.*, 2013). Design-based thinking, or design-based innovation, is a process. Verganti (2009: 4) references the process as:

> People do not buy products but meaning. People use things for profound emotional, psychological and societal reasons, as well as utilitarian ones. Firms should, therefore, look beyond features, functions and performance and understand the real meaning users give to things.

Design-based thinking involves participation, group activity, brainstorming, evaluation and collaboration, similar to the Postma and

Yeoman (2021) constructivism approach, focused on participation, voices of those research, ownership of the problem and co-construction. Design thinking involves 'designers' solving problems through collaborative integrative thinking, using 'abductive' logic, which is what Lub *et al.* (2016: 251) states as:

> Abductive logic is described as the only type of logic to introduce any new idea, which takes place through the process of forming an explanatory hypothesis. In this sense, abductive logic is different from deductive and inductive logics, which are the logics of 'what should be' or 'what is'. Moreover, designers generally reject a linear, sequential, analysis–synthesis–evaluation scheme but rather approach problems in a spiral structure, recognising the importance of prestructures, presuppositions or protomodels as the origin of solutions and follow a conjecture–analysis cycle in which the designer and other participants refine their understanding of both solution and problem in parallel.
>
> In this sense, design thinking as an approach to future scenarios probably resembles most closely the intuitive logics school in scenario building, in that it consists of ongoing learning activity and is an essentially subjective and qualitative approach to the problem. Although design thinking shares some views and methods with scenario forecasting, it probably differs most from scenario forecasting in its scope. Where scenario forecasting commonly departs from broader societal trends, or at least organisations, in relation to their environments (macroscopic perspective) design thinking departs from the individual in his/her environment (microscopic perspective).

5.5.2 The LifeStyle Hub process

The LifeStyle Hub, a prototype for a future function of the hotel, is a concept that was developed during a future scenarios project for the hospitality industry. The main aim of this project was to provide a future vision and input for innovative business concepts to help hospitality entrepreneurs understand and meet future demands. The first step consisted of an extensive trend analysis, which helped the multidisciplinary research group to familiarise themselves with the field and allowed for exploration of new emerging patterns. In the second step, the people deep dive phase, a series of extensive context-mapping interviews were conducted, in order to gain deeper understanding of the future user, the persona. Typically, a persona is described in narrative form. In the third step, future strategic objectives were defined and the results of the trend analysis and the people deep dive were converted during a 4-day workshop into a prototype called the Lifestyle Hub, which contains all the ingredients for a future hotel function.

5.5.3 The creativity

In order to develop scenarios and take workshop participants into the future Lub *et al.* (2016) used cartoon style diagrams, similar to Checkland's (1990) Rich Pictures, in order for participants to visualise and immerse themselves into the futures as 'future tourist's' and imagine future hotels and lifestyles. Figure 5.3 represents the creativity of the project process.

Figure 5.3 Cartoon style diagram – LifeStyle Hub
Source: Lub *et al.* (2016), Sage Publications, reproduced with permission.

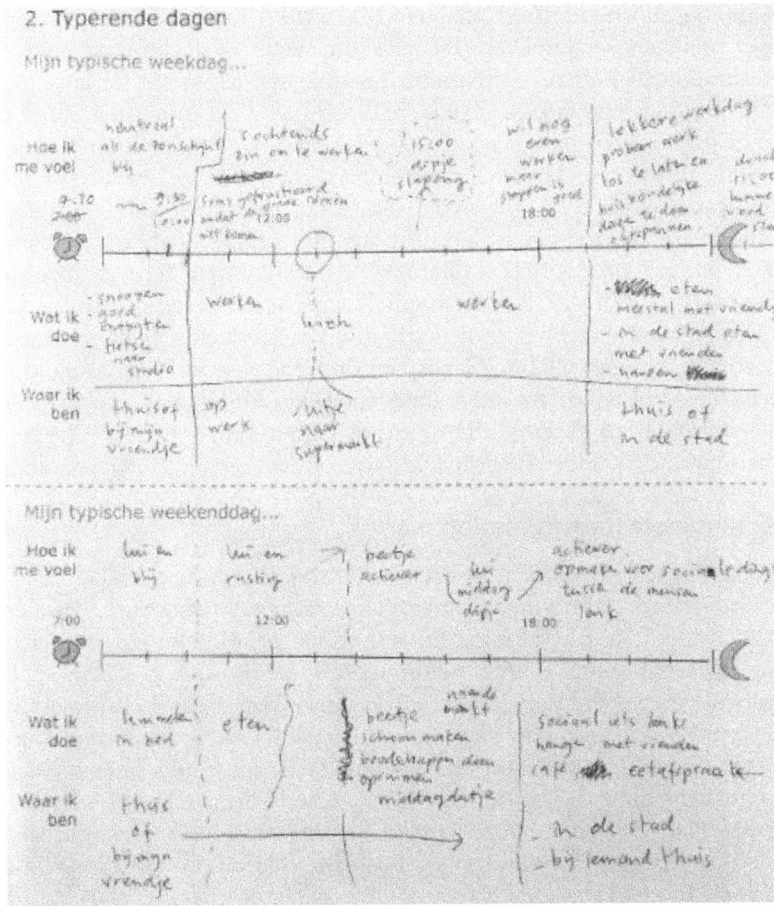

Figure 5.3 Continued

5.6 Case Study Five: Using Scenario Planning to Explore the Future of Work Through Technology

5.6.1 Overview: Is technology a substitute for labour supply?

COVID-19 has accelerated a number of work patterns trends, affecting the fundamentals of how we work, when we work and why we work. Tourism and hospitality is not a 'work from home' industry, it is a service industry that is about the provision of food, accommodation, and related activities. COVID-19 has affected the industry in a number of ways, in particular accelerating trends such as the scarcity of labour (Yeoman et al., 2022). is to explore the future of work by asking the research question 'is technology a substitute for labour supply'. Technology trends including automation, machine learning, brain computer interfaces, molecular nanotechnology and laws of accelerated returns etc., leads us to the concept of technological singularity (Callaghan *et al.*, 2017; Kurzweil, 2005) in which superintelligence completely changes our view of technology and the interfaces with humans. This raises the question about the advancements of technology, the pace of change and technological convergence and the implications on business, including the hospitality and tourism industries.

The future of labour supply has been identified as one of the most important trends that will shape the future of tourism and hospitality industry because of ageing populations and falling birthrates (Hussain, 2022; Mason *et al.*, 2022). Globally, as destinations recover from the impacts of COVID-19, labour supply has become a predominant issue for the leaders and stakeholders of destinations as it is a constraint on capacity (Lugosi & Ndiuini, 2022). This is certainly the case in New Zealand. As such, the Ministry of Business, Innovation and Employment has commissioned a study to examine the future of labour supply in New Zealand's tourism industry, addressing the interfaces of demography and technology.

5.6.2 Industry transformation plan

Labour scarcity in the hospitality and tourism industry in New Zealand was identified as having political and strategic importance (Roberts, 2022). Indeed, it had become an important priority for government to plan for and respond to even before COVID-19 (Boston, 2017). The New Zealand government uses Industry Transformation Plans (ITPs) to address the critical issues or developments which will have a significant impact on the future of the country's economy. ITPs are a high-intensity, high-engagement approach to industry policy with the purpose of setting a transformative vision and action plan for key sectors in the New Zealand economy through the Ministry of Business, Innovation and Employment (MBIE). Tourism has an ITP because of its significance to the New Zealand economy. The ITP for the tourism industry commissioned a research study about the future of employment in the industry. Its purpose was to enable better work and opportunities for those in the tourism and hospitality industry which is a key part of creating a regenerative tourism system.

One of ITP's objectives was to consider the future of work and what is 'better' work. Thus, based upon the expertise of Dr Ian Yeoman at Victoria University of Wellington, a team was brought together led by Dr Yeoman create a set of scenarios about the future of work for New Zealand's tourism industry resulting in the following and research question being established.

5.6.3 Scenario question

In order to shape the future of work, a focused question was used, namely: *is technology a substitute for labour supply?*

This question was addressed by constructing four scenarios about technology and labour supply in 2035. Each scenario represented an alternative future incorporating a scenario summary, signals of the scenarios presently occurring and examples of policy decisions and implications as if the scenario was taking place in the present. The purpose of the scenarios was to paint a picture about the future and for users to work with the scenarios to find answers and actions. They are not a forecast of what 'will' happen but a picture of the future of could happen. Its about inventing and imaging the future (Ball *et al.*, 2017; Sardar, 2010).

5.6.4 Research methods

A scenario planning methodology was deployed that follows Dator's *Alternative Futures* (Dator, 2009) method whereby four scenarios are presented that represent different narratives, circumstances, events and data projections. Alternative Futures is based on the principle of ontological plurality (Yeoman & McMahon-Beattie, 2018) where no one can predict an exact future but several futures can be considered. The four scenarios use the following classification of ontological difference:

- Continuation (business as usual, more of the status quo growth).
- Discipline (behaviours to adapt to growing internal or environmental limits).
- Collapse (system degradation or failure modes as crisis emerges).
- Transformation (new technology, business, or social factors that change the game).

The scenarios were constructed initially using Pierre Wack's *Shell Method* (Chermack & Coons, 2015; Wack, 1985; Yeoman *et al.*, 2022) which is fundamentally a secondary research process or 'kitchen table' method based upon gathering evidence from a range of different sources for scenario construction, then testing the scenarios with key stakeholders for development. Twenty leading stakeholders were identified by the Ministry of Business, Innovation and Employment's (MBIE) New Zealand Strategy tourism secretariat for participation in the research. Stakeholders represented a wide range of sectors in tourism and were key influencers and power brokers. MBIE where the sponsors of the project, as they wanted the scenarios to test future policies related to the New Zealand hospitality and tourism workforce. Three online workshops were organised, each with 20 stakeholders using online Miro boards (Lee, 2019). Each workshop lasted 120 minutes. In workshop 1, trends and drivers of change were identified for impact and uncertainty. In workshop 2, a scenario matrix (see Figure 5.4) was discussed which was the foundation of the scenarios. In workshop 3, implications and policy decisions were discussed.

5.6.5 Overview of scenarios

- **Scenario 1:** *Robbie the Chef* represents a world without human chefs in which production robots run the kitchen. This is a fully automated

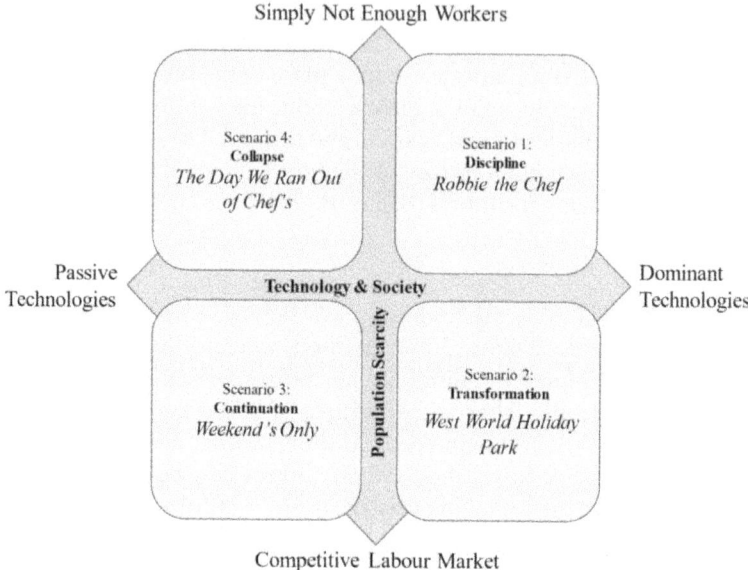

Figure 5.4 Scenario planning matrix – Future of Work

world. This is an example of the dramatic rise of robotics, automation, the advancement of science and rapid changes in technology. This is a fully automated world with a new leisure society. These changes address the massive labour shortages brought about because of an ageing population and competition between sectors. We tell the story of Robbie the Chef®, New Zealand's 3-star Michelin Chef.

- **Scenario 2:** *West World Holiday Park* is a popular tourist attraction about indulgent experiences shaped by advanced robots and avatars. The park was designed by Beta Workshops in Dunedin, a technology startup of Eastern European immigrants who came to New Zealan in 2020. We tell the story of Igor Sikorsky and New Zealand's thriving technology sector and how tourism became an automated and contactless industry.
- **Scenario 3:** *Weekends Only* is a scenario about tourism businesses' constant struggle for labour which resulted in a smaller but more professional industry. We tell the story of hospitality graduate Faizan Ali who moved to New Zealand in 2022 and subsequently started his own award winning restaurant.
- **Scenario 4:** *The Day We Ran Out of Chefs* represents how the tourism and hospitality sector became unsustainable as it couldn't compete with other industries which offered better terms and conditions. We tell the story of a conversation between Catherine and Sheila. Catherine manages the Lodge in Milford Sound. The Lodge offers minimum service with no frills. Where possible, labour-saving technologies are

used. Sheila is a visitor who recalls her experience of the Great Walk and how she used personal technologies to enhance her experience i.e. Department of Conservation (DOC) Extended Really apps.

5.6.6 Scenario implications

New Zealand demographics are startling. Any growth in tourism is going to be constrained by issues of labour capacity. Ultimately, we will run out of young people under present population forecasts (Page, 2022). The key issues that are raised in the scenarios included the need for innovation, higher productivity, increased use of technology and the impact of a changing demography. The ITP panel was asked to consider the scenarios as a complete set, highlighting the key questions for further research and analysis. These questions where:

- Does New Zealand want technology as a substitute for real people or will we have no choice?
- What is the added value of a career in tourism beyond the traditional roles such as chefs, tour guides or casual work?
- How can we use technology to improve productivity without affecting the personal experience?
- At what point in the future does New Zealand's tourism industry become unsustainable because of the scarcity of labour?
- Is the only valid alternative to encouraging New Zealanders to think about a career in tourism a policy of immigration?
- How do you communicate that tourism is the most entrepreneurial industry in New Zealand?
- New careers will be established as technology creates new experiences and products, thus how do we to communicate and educate New Zealanders about these opportunities?
- Technology has revolutionised gaming. What can it do for tourism?

These questions where then further developed by MBIE (2023) in the report He Mahere Tiaki Kaimahi – Better Work Action Plan as outcomes which was published as a consultation document.

5.7 Concluding Remarks

Scenario planning is a process of predicting multiple, plausible and uncertain futures. VisitScotland was the first organisation in tourism to adopt such a process (Yeoman & McMahon-Beattie, 2005: 275).

> Scenario planning is the capability of VisitScotland to perceive what is going on in the business environment, thinking of the consequences of what this means and taking action. The objective is to give Scotland a competitive edge when times get difficult. It is the understanding of the dots on the horizon, perceiving, thinking and taking action in a meaningful way.

This was the beginning that set the prescind for the ETFI and this book. Now, scenario planning is the main foresight methodology used to construct the future of tourism.

It was Chermack *et al.* (2001: 7) who said:

> As the world progresses further into the knowledge age, organisations are faced with an increasing need to respond quickly to a variety of changes. Uncertainty is becoming an important factor for business leaders and planners to consider. In such a rapidly changing business environment, the ability to adapt quickly to major changes can mean the difference between a thriving business and bankruptcy. These changes are often external to the organisation and coping with them has forced managers and executives to adopt a systems view of business. With global complexities and changes likely to continue on the current path of growth, the future of the global business environment will require an even more thorough ability to examine the forces of change and anticipate possible solutions to potential problems.

Twenty-two years later, the world seems even more uncertain and the word VUCA appears in the language of business. VUCA stands for Volatility, Uncertainty, Complexity and Ambiguity (Lubowiecki-Vikuk *et al.*, 2023), whether it is the speed of change associated with AI, how climate change in the Mediterranean is shifting seasons, the future of work shaped by demography trends and continued disruption of war. To many, the way to make sense of the future is through scenario planning and, in this chapter, we have tried to demonstrate some of the uses within the tourism sector, focusing on different reasons for application and different usage. Whether it is understanding the role of politics in destination strategy or how creativity can be applied in the design of hotels.

5.8 Discussion Questions

(1) What are the advantages of using scenario planning to determine the future?
(2) Why is an understanding of organisational politics an important feature of a scenario planning intervention?
(3) What are the advantages and disadvantages of using cartoon style diagrams to represent the future in scenario planning?
(4) Using the trends evaluation matrix in Figure 5.2, apply the matrix for a problem in your business or environment?

Notes

(1) Principles of the New Zealand Labour Party: https://www.labour.org.nz/party_info.
(2) The New Zealand National Party: https://www.national.org.nz/values.
(3) Māori Party: https://www.maoriparty.org.nz/about_us.

6 The European Tourism Futures Institute Method

Learning Points

- Appealing and inspiring scenarios are a combination of words and images.
- Scenario development and strategic foresight are lively and creative processes that take place in moderated sessions with clients and stakeholders.
- Scenario development and strategic foresight require thinking in possibilities, requiring an open mind and outside-the-box thinking.
- Strategic foresight is a cyclic process: because of ongoing changes in the environment, scenarios and strategies need to be adaptive and adjusted on a regular basis.

6.1 The Futures Cone

In the previous chapter, three kinds of scenarios were introduced:

- Explorative scenarios: What *could* happen
- Predictive or extrapolative scenarios: What *will* happen
- Goal-oriented or aspirational scenarios: What *should* happen (normative)

The differences between the three types of scenarios are illustrated by means of the future cone, developed by Voros (Voros, 2017). The preferred future is positioned in the heart of the cone as a goal-oriented scenario. Whether this preferred future can be achieved is partly affected by the probable future as described in a predictive scenario, and partly by the possible future described in multiple explorative scenarios (see Figure 6.1).

6.2 Formulating and Delimiting the Strategic Question

A first and important step in scenario planning is to formulate the strategic question, that is, a question that is surrounded by uncertainty

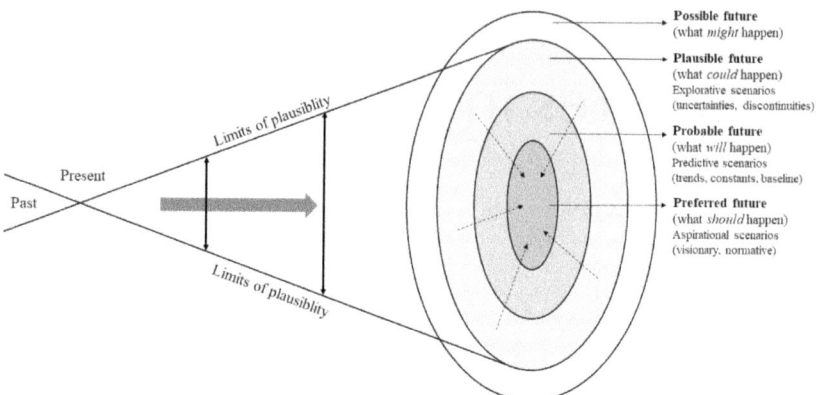

Figure 6.1 Futures cone – three types of scenarios compared
Source: Adapted from Voros (2017).

and is not easy to answer. In the case of a client, this is done during the intake. A proper question to start with is to ask the client what really puzzles the organisation with reference to the future, or, to put it metaphorically, what keeps the decisionmaker awake at night. Another possible question is: what are the future problems or issues that may become crucial for the organisation? The discussion should result in a so-called scenario question that will guide the scenario planning process. An appropriate scenario question is of a strategic nature, addresses the real problem, is goal oriented, is specific, simple and objective, ideally breaks with dominant thinking or taboos and specifies the long-term time horizon that is taken into consideration.

A useful tool to delimit the topic is a domain map. That is a simple mind map with the direct and, if preferred, the indirect aspects of the topic that should guide the scenario process. A listing of primary, secondary and tertiary stakeholders of the strategic question may also be useful. Both the domain map and the stakeholder listing help to prevent blind spots during the development process of the scenarios. The stakeholder map may also help with identifying which stakeholders should be invited in the scenario planning process (Hines & Bishop, 2015).

6.3 Developing a Predictive Scenario

A predictive scenario (see Figure 6.1) is a scenario that describes the single future that is likely to evolve. The first step is to map the context by making an era analysis. Successive periods of relative stability and coherence are mapped out. An era usually begins and ends with a disruptive or otherwise discontinuous event, within or outside tourism (e.g. the banking crisis in 2008, the COVID-19 pandemic that started in 2019, the opening of the first airport on a distant island, the introduction of low-cost airlines or the start of Airbnb) and often has its own identity based on what is dominant in that period ('the era of the …'). By making an era analysis of the domain together with the stakeholders involved, insight is gained into the period in which the domain is positioned at the present, which disruption occurred that initiated the period and the circumstances and stakeholders that play a role in it (Hines & Bishop, 2015).

The results of the era analysis form the basis for the next phase: the structural analysis of the domain in the current era. This primarily focuses on understanding how the domain works, that is, what the most important variables and stakeholders are that influence the functioning of the domain in the current era. Second, it is important to know how the domain currently performs based on the values of the important variables. In the stakeholder analysis it is important to know the extent of the influence of the various stakeholders (Hines & Bishop, 2015). First, the stakeholders need to be listed, after which their role and power can be analysed using, for example, the power × interest matrix by Mendelow (1991) or the Salience Model by Mitchell et al. (1997). In the Salience Model three variables are used (power, legitimacy and urgency) to define seven categories of stakeholders: dormant, discretionary, demanding, dominant, dangerous, dependent and definitive stakeholders. In short, the structural analysis represents an analytical snapshot of the present, that helps to understand the circumstances and performance of the current era.

When a predictive scenario is developed, the snapshot of the present is, as it were, slowly set in motion until the end of the period of investigation. In order to set the snapshot in motion, the factors and variables need to be known that drive the surprise free and linear development of the domain within the present era. Therefore, desk research needs to be conducted on (Hines & Bishop, 2015):

(1) Trends. This refers to the gradual or stepwise development of a quantity over an extended period in a particular direction.
(2) Forecasts of specific quantities and their value at a particular point in time in the future.
(3) Constants. Conditions that are not expected to change within the time frame examined.
(4) Cycles. Qualitative or quantitative changes in the domain that recur within the period under study.
(5) Plans. Intentions or plans of key stakeholders that are announced or in the pipeline, and with which they want to initiate change within the time frame studied. Plans that were announced before but will be discontinued are also included.
(6) Projections. Public predictions that may influence people's behaviour in case of a special event.

After the structure of the domain in the current era has been mapped out, and the factors that drive its future development in foreseeable direction, the data will be reviewed, interpreted and analysed. It is key to identify the most powerful attributes that drive the developments in either a stimulating or inhibiting manner, to extrapolate their values and their synergetic effect. It is also recommended to consider the possible discontinuing force that could cause the era to end. Next, the analysis results in an outline of the most likely future of the domain by the end of the study period: the predictive scenario.

6.4 From Baseline to Alternative Scenarios

The predictive scenarios can be used as a baseline scenario that portrays the probable future (see Figure 6.1). Since it is acknowledged that this baseline is almost always wrong (Hines & Bishop, 2015), alternative futures can be explored by speculating about how incorrect it may be. This is done by reviewing the assumptions of the baseline scenario one by one, to ask critical questions about each assumption (is it true? is it relevant? is there sufficient justification?), and, in case of doubt, to reverse these assumptions by assuming the opposite, if plausible, and to use the 'reversed' assumptions to create alternative images of the future. Alternatively, potential wild cards or disruptions and their positive and negative implications could be explored to create alternative scenarios (so-called *wild card* scenarios). This working method is rooted in the idea that the future will be a result of the baseline scenario in combination with surprising developments leading to alternative possibilities. It is common that baseline and alternative scenarios are developed with a relatively small team of experts. The outcomes may be used to inform existing policies and strategies or develop new strategic courses of action, which refers to strategic foresight (Hines & Bishop, 2015).

6.5 Developing Explorative Scenarios

The aforementioned approach, in which a baseline scenario and alternative scenarios are constructed, has been applied for NBTC Holland Marketing in 2018–2019 (see Chapter 10, Scenarios for Inbound Tourism to the Netherlands). In most cases, ETFI practices the scenario planning and foresight process as illustrated in Figure 6.2. The first phase is about constructing the base. In fact, a model is created that represents the current state of the system – the subject under study and its environment – and its dynamics. It is about defining the key variables and how they are related. The second phase focuses on the construction of the scenario framework. It starts with analysing the problem and formulating a clear scenario question. During the third phase the (explorative)

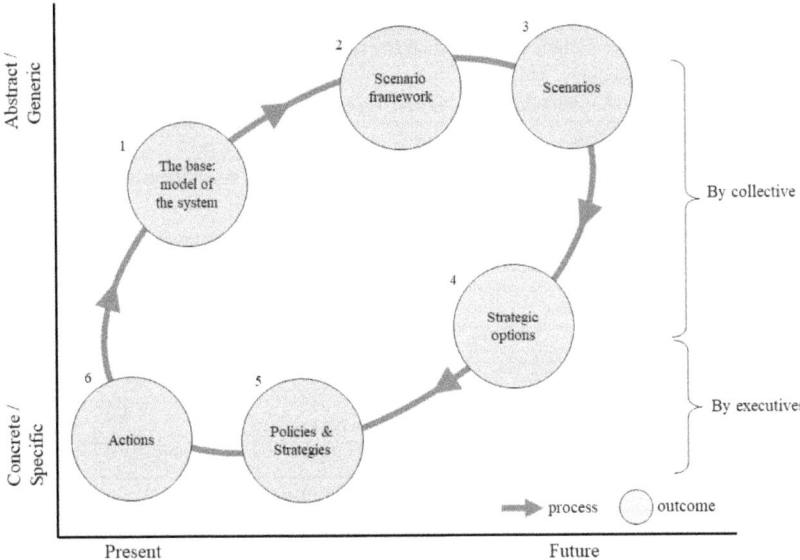

Figure 6.2 Strategic foresight process

scenarios are crafted. An evaluation of the opportunities and threats of the alternative scenarios may be used as a source of inspiration to develop a new vision or business model (aspirational scenario). Alternatively, the evaluation of opportunities and challenges of the explorative scenarios could also be transferred into policy domains and subsequently into strategic courses of action, which may entail new concepts, products or services. At any time, it is recommended to institutionalise strategic thinking by structurally embedding strategic foresight in the organisation and establishing a system of coordination and a system of horizon scanning. Finally, these actions are put into practice, in order to reach the strategic objectives, which is a step that the ETFI is not involved with. This is also the start of a new cycle. As the contextual environment of a business or organisation is dynamic and uncertain, scenario planning and strategic foresight constitute an ongoing process.

Questions during the phases of the strategic foresight cycle are as follows:

(1) Who am I? (The prerequisite question, too often ignored, strengths and weaknesses.)
(2) What could happen? (Scenarios.)
(3) What can I do? (Strategic options.)
(4) What will I do? (Strategic decisions.)
(5) How will I do it? (Actions and operational plans, and an essential prerequisite question.)

Table 6.1 Steps in the adaptive strategic foresight process

Preparatory phase: Formulating and delimiting the scenario question
Phase 1. Horizon scanning
Phase 2. Environmental analysis (patterns, processes, driving forces of change)
The base: model of the (complex adaptive) system
Phase 3. Limits of the plausible, importance by uncertainty matrix, key certainties, key uncertainties
Phase 4. Establish scenario cross (or triangle)
Scenario framework (scenario cross)
Phase 5. Scenario foundations: associations and ideas to fill the scenarios
Phase 6. Scenario development
Scenarios
Phase 7. Scenario implications
Phase 8. Opportunities and threats of the scenarios
Phase 9. Strategic options that bridging opportunities and threats
Strategic options
Phase 10. Ex ante evaluation of strategic options and prioritisation
Policies and strategies

In this approach, complex adaptive systems thinking is key. Eventually, alternative pictures are crafted of how we think the system could have evolved after a number of years. The process starts with collecting data about the system (horizon scanning, future scanning), with a focus on data that are expected to have future relevance. Next, the data are qualitatively analysed. Patterns and processes are deduced with the aim to understand how the systems works and to identify the factors that drive the change of the system. Grasping the dynamics of the system helps to paint a picture of how it could have changed by the end of the period of investigation in a plausible way. The steps to develop explorative scenarios and policies and strategies that are inspired by these scenarios are described in more detail below (see Table 6.1). Ideally, most phases are conducted in moderated group sessions with the client, a clients' staff team (such as the management team and/or the marketing team) and any relevant stakeholders. During these sessions it is the facilitator's task to ensure that all participants are heard and are able to contribute to the discussion. Part of the strategic learning process is the discussion between the participants and exchange of arguments that eventually lead to consensus. The identification of policies and strategies is preferably done by a smaller team of executives, due to confidential characteristics. During the COVID-19 pandemic, the ETFI put a lot of effort into holding moderated sessions online. MURAL proved to be a valuable software solution for this.

Figure 6.3 Visualisation of the outcome of horizon scanning

6.5.1 Phase 1: Horizon scanning

Horizon scanning (also referred to as future scanning or environmental scanning) is about spotting early indications of developments that may become more important over the course of the research period. These may be indications of the changing power position of stakeholders, of unsolvable or seemingly unsolvable conflicts, of new plans or existing plans that may not go ahead, of innovative ideas, of new trends or counter trends or of possible game changers or disruptive forces. The indications of imminent developments may relate to the macro-environment (demographic, economic, social, technological, ecological or political developments), to the meso-environment (in the domain, sector, market) or to the micro-environment (such as, specific services, products, start-ups, innovations, behaviour of tourists) (Postma & Papp, 2021). All observations are written down on separate sticky notes (see Figures 6.3 and 6.4). For the purpose of the horizon scanning, a multitude and great diversity of sources must be consulted or, preferably, monitored continually (see Table 6.2, for example, Hiltunen, 2008).

The ETFI has been using two workshop approaches successfully. With the first approach, the participants come to the workshop with an open mind, and they need to rely on their prior knowledge. With the alternative approach, the participants are asked to do some preparatory work on a semi-structured *pro forma* with instructions that they receive about one or two weeks before the session. The participants can use the

Figure 6.4 Example horizon scan with four representatives from city DMOs in Europe

pro forma to register observations and ideas to be taken to the session. Ideally, each observation meets the following two criteria: it is relevant for the scenario question and it points at an anticipated change or development. In some cases, ETFI staff also contributes to the horizon scanning. The workshop horizon scanning is normally scheduled in conjunction with the workshop environmental analysis.

Table 6.2 Possible sources of horizon scanning

Newspapers	Newsletters	Think tanks
Websites	Magazines	Futurists
Blogs and vlogs	Books	Work by artists
Wikis	Book reviews	Books by science fiction writers
Library databases	Presentations (e.g. ITB)	Films
Podcasts	Reports	Scientific reports and sites
Videos	Studies	Consultants
News sites	Interviews	Researchers
Chat rooms	Seminars	Sociologists
Trends watchers	Management gurus	White papers
Philosophers	Conference announcements	
Technical papers		

6.5.2 Phase 2: Environmental analysis

Environmental analysis is also referred to as Causal Layered Analysis, in short CLA (Inayatullah, 1998). The many observations from the previous step are clustered on the basis of mutual coherence (see Figure 6.5). This primarily concerns cause–effect relationships. Observations that evoke the same feeling or mean the same thing can also be grouped into the clusters. The clusters should be internally consistent and sufficiently different from each other, without overlap. If necessary, clusters should be split and/or merged. Ultimately, each cluster represents a process that influences the development of the domain in the agreed time frame. The final step during the stage of environmental analysis is to determine the driving force of each of these processes. It is the fuel that keeps the process running. The environmental analysis can result in driving forces in the macro- or contextual environment, in the transactional or meso-environment and/or in the own business or organisation.

The workshop horizon scanning and environmental analysis can be done during the same facilitated session. The participants may be overwhelmed by the large number of sticky notes. It may go up to 100 or more, such as illustrated in Figure 6.4. To start the procedure, it is suggested that someone starts with finding two observations that are perceived to be causally related, takes them up, puts them together on the wall and explains how and why they impact upon each other. If there is consensus, another observation may be added to the cluster, the process is repeated and the cluster grows. Arrows may be added with a

Figure 6.5 Visualisation of clustering of the outcomes of the horizon scan into key processes

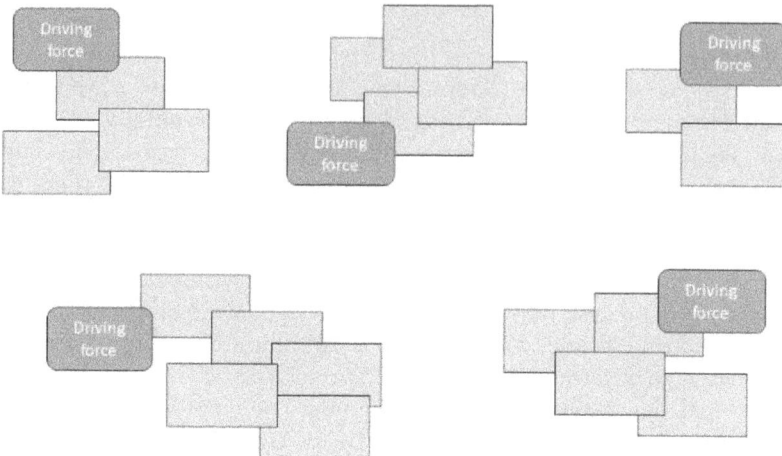

Figure 6.6 Visualisation of driving forces identification

marker to underline the perceived causality or correlation. To prevent it from getting too big, eventually a new cluster should be shaped in the same way. The clustering process is repeated until all observations are combined in clusters, except the ones that do not seem to belong anywhere. During the clustering process, discussions may lead to modifications of clusters. Once the clustering is finished, the clusters need to be reviewed for internal consistency and overlap and, if necessary, they need to be adjusted. Finally, cluster by cluster the process is reviewed centrally in such a way that all participants are able to explain the process occurring in each cluster.

Collaboratively, a driving force is identified that fuels the process in the cluster (see Figure 6.6). The driving force may be at a higher level of abstraction than the sticky notes in the clusters, but it may also be that one of the existing sticky notes can be regarded as the driving force. Beware that the label of the cluster is not a theme, but a force that points at change and drives the change. Eventually the process leads to collaborative understanding of the relevant part of the external environment and how it operates. In fact, the identified driving forces and the processes that they fuel represent a simplified model of the societal system as it is perceived by the participants.

6.5.3 Phase 3: Interpret the dynamics of the driving forces

The first step in the interpretation of the driving forces is to explore in which two extreme, but plausible, directions the underlying process of each force could develop during the study period. These extremes define

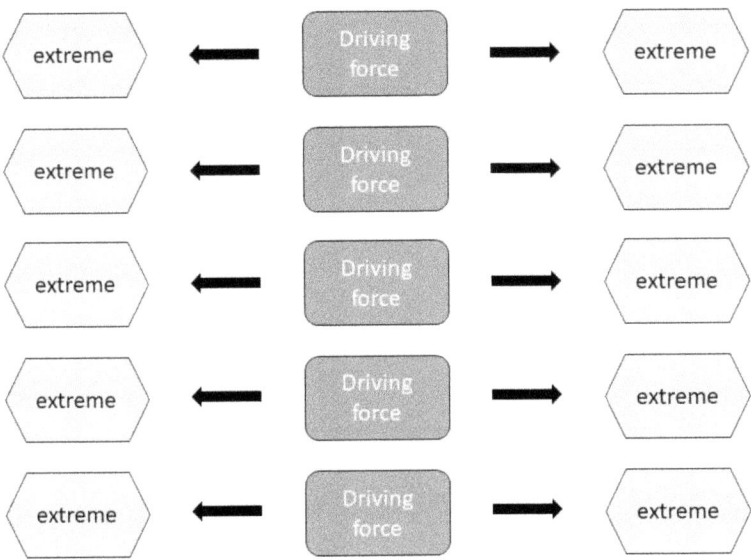

Figure 6.7 Visualisation of driving force's limits of the plausible

the 'limits of the plausible' (see Figure 6.7). These extremes could be quantitative (e.g. gas provision should either lead to a shortage of gas or an abundance of gas) but qualitative opposites give more inspirational scenarios at a later stage (e.g. gas provision could lead to a hydrogen economy or a solar and wind economy). Defining adequate extremes is an important crux of the entire scenario development process. It requires creativity and imagination. To challenge the dominant thinking, the participants need to discuss the processes in depth, exchange views and arguments and challenge each other's thinking.

If the extremes are set, a new round may be needed to make each extreme even more extreme (yet plausible). The second step at this stage of interpretation is to position the driving forces in an importance by uncertainty matrix. This means that the driving forces are ranked relative to each other according to the degree of impact on the domain and the degree of unpredictability/uncertainty. The result is a kind of scatter plot, with the degree of unpredictability along the horizontal axis (from low at the left to high at the right) and the degree of impact along the vertical axis (from high at the top to low at the bottom). The bottom part of the plot shows the driving forces that are of secondary importance. They can be disregarded during the rest of the scenario planning process. At the top left are the driving forces that are of great influence and are relatively predictable or certain in their future development. We can consider them as the key certainties.

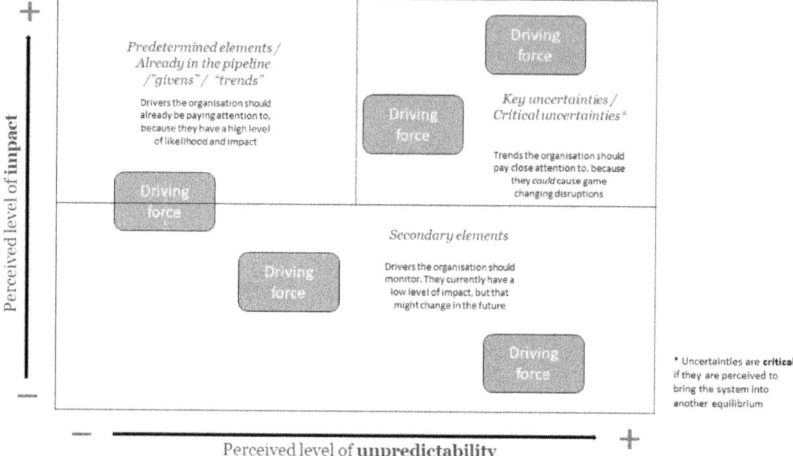

Figure 6.8 Visualisation of impact by uncertainty matrix

At the top right are the *key uncertainties*. These are the driving forces that are expected to have a major influence, but whose development is extremely uncertain. The core uncertainties can be considered as critical uncertainties if it is plausible that they could transform the entire system and move it to another equilibrium. So, in fact, such critical uncertainties are potentially disruptive.

In projects conducted by the ETFI the phase in which the driving forces are compared with each other concerning their level of impact and level of uncertainty, has been done in two ways, depending on the size of the group of participants. Whichever of the two approaches is used, the outcome is an importance by uncertainty matrix as in Figure 6.8. To create explorative scenarios two key (or critical) uncertainties are the starting point. These driving forces have a strong impact on the future of the part of the environment that is relevant for the scenario question, while the way in which these forces drive the future is quite unpredictable. Either way, firstly each driving force has to be written on a separate sticky note first.

If the group of participants is relatively small, participants can be asked to place the sticky notes vertically on top of each other in order of degree of impact. The driver with the greatest impact at the top and the driver with the least impact at the bottom (see Figure 6.9a). Next the participants are asked to keep the sticky note at the vertical position, and slide them to the left or right, depending on the level of uncertainty/unpredictability. The most certain/predictable at the far left; the most uncertain/unpredictable at the far right (see Figure 6.9b). Both the vertical and the horizontal positioning is a group effort and requires discussion by the group members and the exchange of arguments.

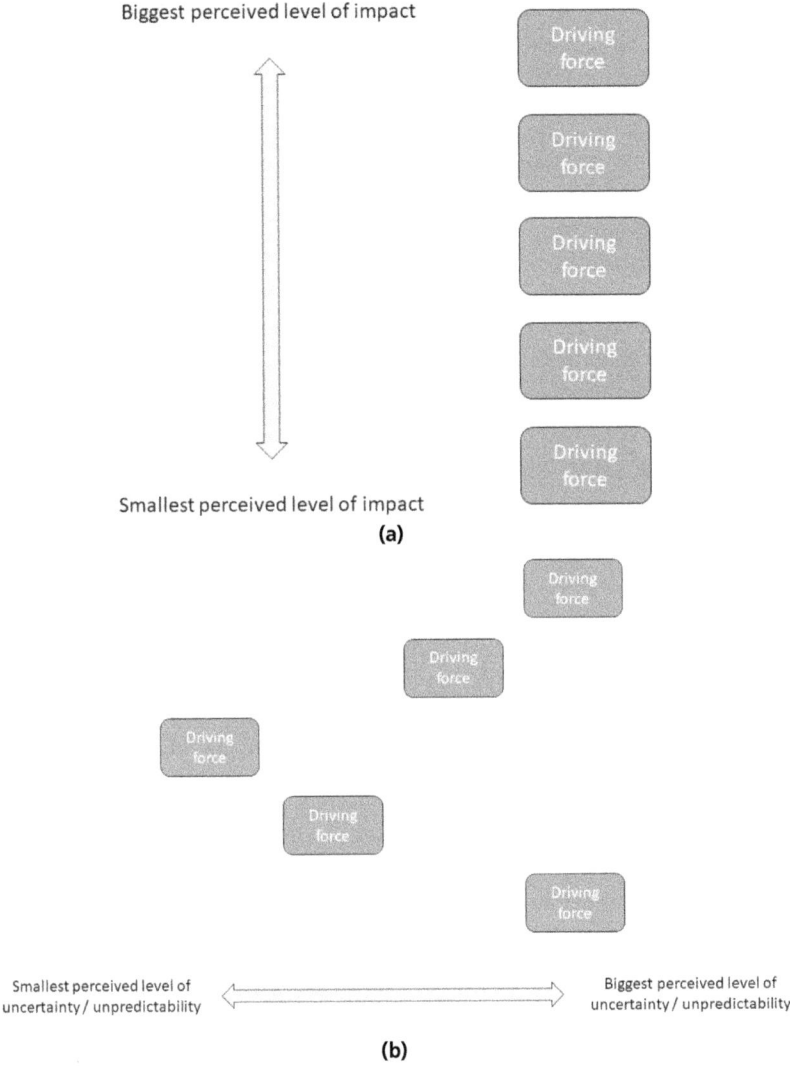

Figure 6.9 a,b Visualisation of the shifting and sorting of the driving forces

If the group of participants is fairly large, the best approach is to use dot voting. All participants receive three stickers in colour A (the number of stickers may be less or more). Each participant is asked to select three driving forces to put their sticker on. Next, the participant receives the same number of stickers in another colour, and they are asked to select three drivers of which they think are most uncertain/unpredictable. The number of stickers of both colours is used to position the sticky notes with the driving forces into the importance by uncertainty matrix.

With either approach, eventually two key uncertainties have to be chosen as the two dimensions on which the scenarios will be built. These two key uncertainties need to be independent. That means that they are not both affected by a third driver. In that case, the clustering in phase two was done inadequately and the two key uncertainties should be merged. The driving forces that have not been chosen as dimensions for the scenario cross will not be disregarded but will be used to fill the scenarios with content.

Please note that phases 1 to 3 could also be used to develop predictive scenarios. The givens that end up in the left upper corner of the importance by uncertainty matrix (see Figure 6.8) provide valuable input next to the trends, forecasts, constants, cycles, plans and projections as discussed in Chapter 6.3.

6.5.4 Phase 4: Establish the scenario cross

The two key uncertainties that were selected form the basis for the exploratory scenarios to be developed. When we place those two dimensions with the extremes as identified in phase 3 perpendicular to each other, a cross of axes is created, which is referred to as a scenario cross. When both key uncertainties are based in the macro- or meso-environment, it results in four so-called environmental exploratory scenarios. When both key uncertainties are internal and represent for example two dilemma's, the scenario cross shows four internal also called strategic explorative scenarios. When one key uncertainty is from the macro- or meso-environment and the other internally, the framework includes four so-called system scenarios that make a direct connection between the organisation and its environment (see Figure 6.10).

It is worth mentioning that explorative system scenarios can also be based on a trilemma (see Figure 6.9). In 2004, Shell used this approach. The company listed three alternatives policy options that are all equally (un)attractive and presented them in a trilemma triangle. The trilemma is based on three driving forces in the business environment and each

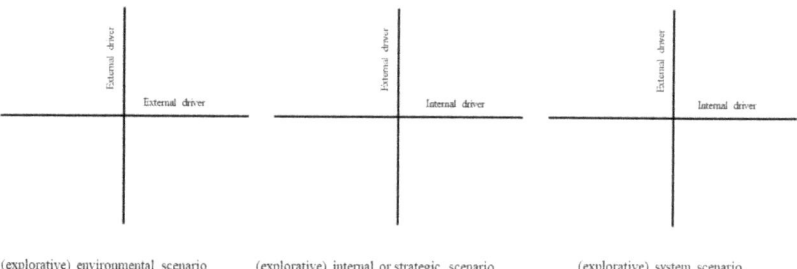

Figure 6.10 Three types of explorative scenarios

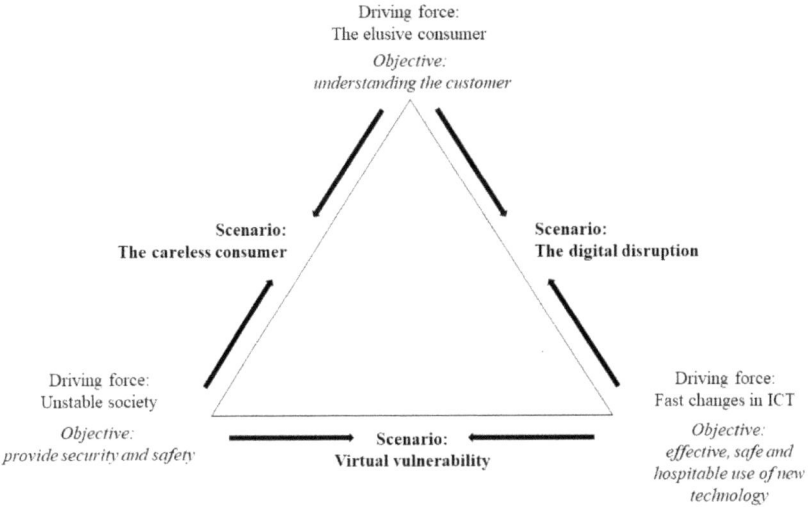

Figure 6.11 Example of a trilemma triangle

is linked to a specific objective. These goals do not go hand in hand, but neither can they exist separately from each other. In the trilemma triangle three scenarios are presented. Each scenario unites two driving forces/objectives and describes the trade-offs. In a project by CELTH, the ETFI has used this approach to establish a trend vision in 2017 as addendum to the annual Trendrapport toerisme, recreatie en vrijetijd [Trend report tourism, recreation and leisure] during three trend panels with CEOs from the industry, entrepreneurs from the industry and with researchers. This trilemma triangle is presented in Figure 6.11 (CELTH, 2017).

In the above description of this phase complex reality is eventually reduced to only two dimensions with two extreme 'scores'. This is common practice, although critiqued by authors, such as Godet and Durance (2011). It should be noted that more than two driving forces may be used to create scenarios, along with two or more scores of each. Every possible combination of scores on all driving forces leads to different scenario. To create such so-called morphological scenarios is a complex exercise for which software is available. Some of the scenarios will not make sense, because specific combinations of scores are simply not possible. These can be crossed out.

6.5.5 Phase 5: Scenario foundations

In the previous phase a scenario cross was created. The scenario cross is the framework of four scenarios. Each scenario is framed by

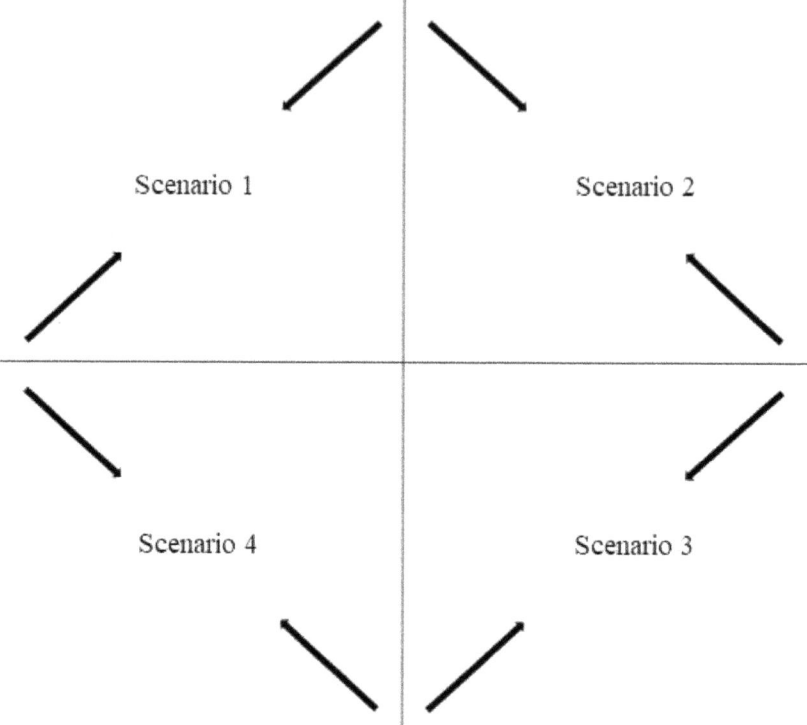

Figure 6.12 Scenarios are framed by the extremes of the two axes in the scenario cross

the combination of two extremes (see Figure 6.12). As the cross is still entirely empty, it first needs to be filled with ideas. To fill a scenario with ideas is not as easy as it seems. Scenarios should be outside-the-box, they should exceed the borders of our imagination, and break out our dominant thinking. Therefore, it is best to focus the mind on one scenario at a time and to have a sufficient break in between working on two scenarios.

Elaboration of the scenarios can be done according to the following steps, below:

(1) Consider the extremes of the driving forces in phase 2 that have not been used for the two axes of the scenario cross. If you think one or more extremes fit with the scenario, write it down.
(2) Close your eyes, use your imagination and try to visualise a scenario that is framed by the two extremes. Let your thoughts flow and dive into this world. What do you see? What do you feel? What do you think? Write down each and every thought.

(3) Turn to the 'key certainties', the givens that were positioned in the left upper corner in the importance by uncertainty matrix in phase 3. These givens are perceived to remain in the future, no matter which of the scenarios will evolve. However, the way to which the givens are expressed in the four different contexts may be different. for each of these givens, try to imagine how, and write down your thoughts.

In projected conducted by the ETFI, the development of the scenario cross and the establishment of the foundations for the scenarios are usually covered in one workshop.

6.5.6 Phase 6: Scenario development

Once the scenarios are all filled with thoughts, feelings and associations they need to be revisited in an attempt to find a title for each that represents its core and goes well with the other titles. The titles could be telegram-style catch phrases, although representative metaphors could also be used. For example, the four names of animals, events, cities, etc. Once the scenarios have a title, new ideas may pop up to be added to the content of the scenarios.

6.5.6.1 The structure of a scenario narrative

The next phase is to connect the associations that were written up in a scenario into a consistent narrative. There are two ways to compose these narratives: structured or unstructured. Structured means that a number of key ingredients are described in each of the scenario narratives. This allows for a direct comparison of the scenarios. Unstructured means that the narrative is inductively created based on the associations that were written down in the scenario.

Either way, the narratives could be written from the distant perspective of the observer, but they can also be written in the first person (I perspective), such as a customer or a staff member. To write an appealing story from a personal perspective, specific storytelling guidelines may be followed. There should be a clear topic, a main character and location, a plot with a linear set of events and one or more plot points (inflection points) that explain how the future is logically developed from the past.

In any case, the scenario narrative should be positioned in the future (the end of the period under investigation) and written in the present tense. The events that eventually lead to that future are written in the past tense. In ETFI practice we recommend starting each scenario with, for example: 'It is 2035'. Besides, each scenario should each be value neutral and give an objective observation of what will happen. That means that persuasive language should be prevented. Which scenario is more preferable, or more likely than the other may differ between stakeholders.

Since the end of 2022, generative Artificial Intelligence (AI) such as ChatGPT has become available to assist with the creation of an appealing narrative of a scenario. The ETFI has experimented with creating text narratives based on clues generated by workshop participants. However, the writing style of these ChatGPT generated stories appeared to be too far away from what ETFI staff would write themselves. Therefore, this AI generated content was only used to get some inspiration for the narrative written by the staff. Besides, the clues were also given to MidJourney in an attempt to create a visual of the scenario narrative for a Dutch National Park. Small tweaks of the text resulted in different visualisations. Afterall a selection was made of the most appropriate one, which is shown below

6.5.6.2 Other requirements of a scenario narrative

Since scenarios are not a goal in themselves but serve as the means to inspire strategic decision making, the scenarios should have no overlap, go beyond dominant thinking and, therefore, surprising, appealing and plausible by the end of the period under investigation. A plausible scenario presents a future that is imageable or an imageable and logical outcome of a consistent chain of events.

Scenarios are communicated best if they activate both halves of the brains: the 'analytical' left half of the brains and the 'creative' right half of the brains. To fully activate the brains of and trigger reactions of the users or their constituency, the scenario narratives are ideally presented together with some form of images. The combination of words and images may take different forms, such as the following examples:

- Photoshopped images with Adobe Photoshop.

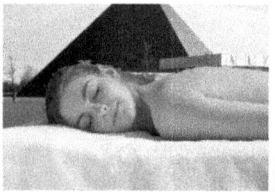  

- Comic drawings:
  see also Figure 6.13:

- Abstract visualisations:
  see Figure 6.14.
- Video clips:

  Four scenario clips made by students

  Four scenario clips made by professionals

- A (digital) magazine that may contain combinations of the above options.

Alternatively, scenarios can also be communicated by means of scenario rooms decorated in the atmosphere of each scenario with discussion tables and note pads, or by means of plays performed by actors.

In the ETFI's practice, several modes of working have been used, as a results of the clients' needs and wishes, and the ideas brought in by the ETFI. Some of these ideas have been applied in student projects, others have been used at conferences and in projects. An approach that the ETFI has applied quite often is collaboration with a bureau specialised in making (comic) drawings. In a number of projects, employees from such a bureau were already involved in the previous phases of the scenario

Figure 6.13 Scenario 'Natural Wadden Sea Dike' in a picture (made by JAM Visual thinking) (project 2015)
Source: Hartman et al. (2014); pictures made by Jam Visual Thinking.

development process. During the sessions they 'took notes' of what the participants put forward in the form of dozens of quickly drawn comic drawings. After the session these pictures were combined into a professionally presented master drawing (see Figure 6.13). In other cases, the bureau was asked to represent the scenario straight away into a picture (see Figure 6.14). In one specific projects (see also Chapter 8), the 'static' picture was used to create a video clip with a voice over and some sound.

6.6 From Scenarios to Strategy

In ETFIs practice, scenarios have been used in different ways in relation to strategy, as follows:

(1) As a source of inspiration to develop a new vision, that is, a (normative) aspirational scenario.
(2) As a source of inspiration to design new strategic courses of action, such as new business, new concepts, new business models, new policies.
(3) As a 'wind tunnel' to test existing (*ex poste* evaluation) or newly developed (*ex-ante* evaluation) policies, strategies or other ideas for robustness.
(4) As a learning tool and frame of reference, to challenge existing paradigms and assumptions and creating a shared perspective or vision to the future.

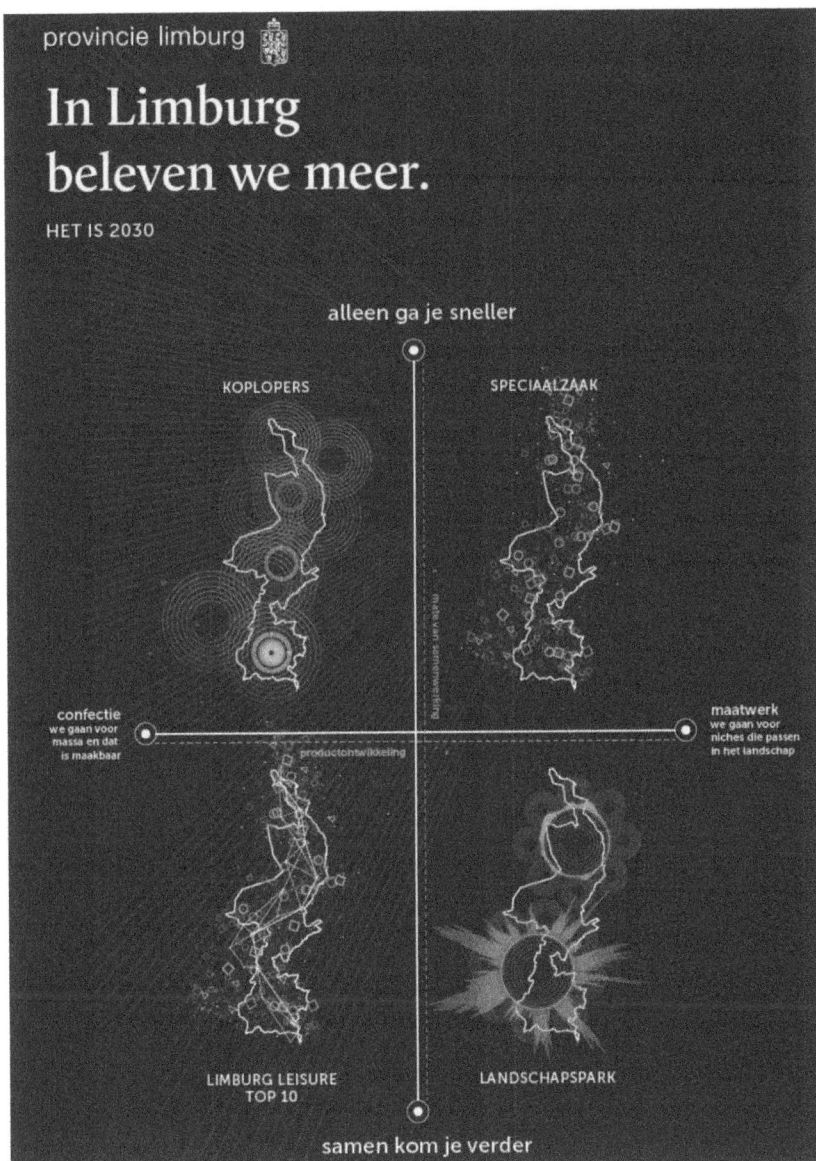

Figure 6.14 Four scenarios for tourism in the province of Limburg, Netherlands, in pictures
Source: ETFI (2018); pictures made by WE ARE KACE.

The explorative scenarios can be used as a source of inspiration to map the needs, wants and ambitions of the business or organisation involved. This could relate to one or more of the scenarios developed. With back casting by means of a futures wheel, stepwise the conditions can be mapped that area required to achieve the desired future as described in the aspirational scenario (see Figure 6.14).

In the remainder of this chapter, it will be explained how strategies can be used to develop new policy or new strategies how the new strategies can be wind-tunnel tested for robustness, in order to pass on the best options to anticipate the future and make the destination, organisation or business more robust. All the steps described below take place during a workshop with the clients and its staff and ideally its stakeholders, facilitated by the ETFI.

6.6.1 Phase 7: Scenario implications

The first step in the development of new strategies is to consider the implications of each scenario in the form of an implication tree (see Figure 6.15) on a big poster with sticky notes on the wall (or on a dedicated MURAL template). In an implication tree, the three key essentials of the scenario are summarised on three separate sticky notes in the middle of the wheel. Since the scenarios describe the macro- and/or meso-environment of the business or organisation, the three key features of the scenario also point at the environment! Because each of the scenarios described an alternative extreme future, there should not be any overlap between the key features across the scenarios.

Next, the challenge is to think from the outside-in and identify three direct implications (direct consequences) of each of those key features

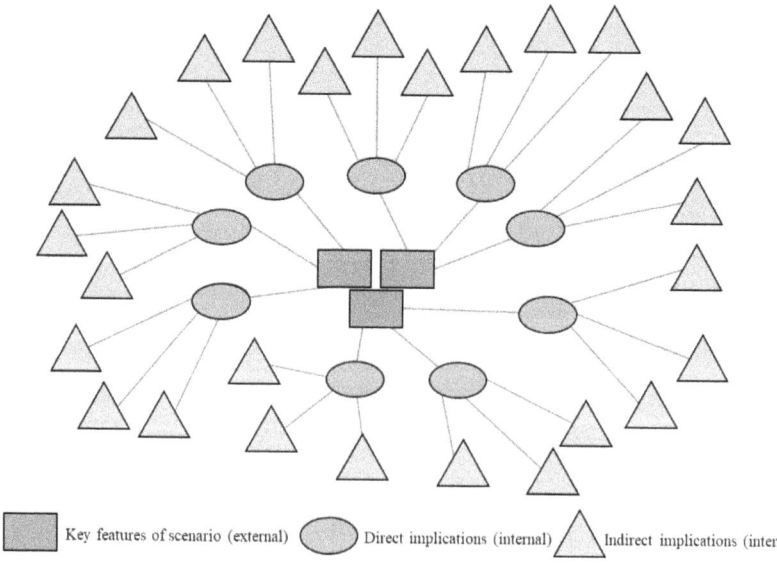

Figure 6.15 Implication tree (forward thinking) or futures wheel (backward thinking)

for the organisation or business in question. Once the direct implications are written up, three implications of the first order implications need to be identified: the so-called indirect or second order implications. The first big challenge in the construction of these implication trees is to think and write swiftly and to not discuss the ideas with the others in the group. The second challenge is to think outside-the-box, to let your thoughts flow freely and to not be inhibited by the present and what may not be possible. Therefore, it is advised to work on one implication tree at the time. The third challenge is to not confuse implications with what the participants desire. Here, it is not about what the participants want or do not want, but what could possibly happen if the scenario would have come true. What would be the direct and indirect implications for the business or organisation in that case? The outcome of this process is four implication trees as in Figure 6.13. Please note, the number of three features, three direct implications and three indirect implications, is arbitrary and may be changed, but works well in practice.

6.6.2 Phase 8: Highlighting the opportunities and threats of the future

Once all the implication trees are ready, it is time to move to what the group would like to have achieved by the end of the period under investigation. What are the implications that the group likes and the ones they do not like? Or in other words: what are the opportunities and threats that each of the alternative futures holds? If the group is not too big, the let's say five to seven, likes and dislikes in each scenario can be chosen based on a facilitated discussion between the participants. If the group is bigger, the alternative is to ask every participant to put three, four or five green stickers on the implications that he or she likes most. Once everyone is finished, the participants are asked to put the same number of stickers on the implications disliked most. It is not allowed for a person to put more than one green or red sticker on one sticky note. Based on the number of green and red stickers, it is easy to identify the most important likes and dislikes of each scenario. It should be notes that the ETFI sometimes only focuses on the likes, the opportunities, to keep a positive drive. The outcome of this step is an overview of likes (opportunities) and dislikes (threats) of each scenario. It is recommended to give each sticky note an identifier code which is useful for later steps. An identifier code could be the number of the scenario, followed by an O (opportunity) or T (threat) and a unique number. For example: S2O6.

6.6.3 Phase 9: Strategic options bridging opportunities and treats

The next step is a brainstorming exercise. The overview of likes and dislikes in all scenarios acts as a source of inspiration to generate ideas of measures or actions to be taken to achieve the opportunities and prevent the threats. The challenge here is to find ideas that combine multiple opportunities and/or multiple threats within, or ideally across, the scenarios. Each idea is to be written on a sticky note together with the identifier codes of the opportunities and/or threats it is based on. Again, it is important to think with an open mind and outside-the-box, in possibilities.

Once the brainstorm of dozens of actions is finished, the ideas on the sticky notes are clustered into coherent strategic domains. Based on the contents of the cluster, each cluster is labelled with a 'strategic course of action' that represents the underlying measures. It could be a new business or new business model, a new concept, a new strategy, etc.

6.6.4 Phase 10: *Ex ante* evaluation of the strategies

In order to formulate a final policy or strategy the strategic courses of actions need to be prioritised. The ETFI commonly prioritises the strategies according to the following criteria:

- Robustness: how many of the scenarios does the strategy anticipate? The identifier codes written down on the sticky notes in each cluster help to find the answer to this question. A strategy that prepares for all scenarios is a robust strategy. If it anticipates just one scenario it is a betting strategy. In case of two or three scenarios, it is called a semi-robust strategy. In most cases it is recommended to be prepared for any future. In some cases, businesses have a betting strategy in place for more than one strategy. Depending on whether the scenarios evolve or not, such betting strategies can be quickly upscaled or downscaled. This activity is also referred to as wind-tunnel testing.
- Suitability: to what extent does the strategic/policy option address the most important issues related to the strategic position of the business or organisation? (Refer to the domain map.)
- Acceptability: to what extent does the strategic/policy option align with the aims of the organisation, and to what extent is it acceptable to the stakeholders?
- Feasibility: to what extent could the strategic/policy option be put into practice, in terms of resources, aptitude and abilities?
- Risk: to what extent does the strategic/policy option address future uncertainties that can affect the organisation or its strategic position negatively, or in other words: how likely is it that something goes wrong?
- Scalability: to what extent is it possible to up or downscale the strategic/policy option quickly with the available personal and financial resources if market circumstances (such as, market demand, competition) require so?

- Sustainability: to what extent does the strategic/policy option address one or more of the 17 Sustainable Development Goals?

The seven criteria above help to make a priority list of the strategic courses of action that were identified. A simple score could be used for each of the criteria from one to four (beware that the risk criterion has a reversed scale), that can be summed up to deduce final priorities. If needed the factors could be weighed.

6.7 Concluding Remarks

Since the ETFI started its operations in 2010, it has come across a wealth of literature on scenario planning and strategic foresight. Much of this literature shows commonalities, but it is also clear that in many cases consultancy firms and authors have tried to give their own twist by modifying or adding specific features or even using different words to express the same. Based on these insights and courses provided by Future Consult in the Netherlands (Jan Nekkers) and the University of Houston (Andy Hines and Peter Bishop), combined with experts' knowledge on complex adaptive systems and resilience in its staff team, the ETFI has gradually developed its own approach of scenario planning and strategic foresight.

The approach, as described in this chapter, reflects of how the ETFI applies the subsequent steps in interactive workshops. The COVID-19 pandemic required staff to transfer the approach to an online setting. A comparable analysis of various software packages resulted in two options that could do the job: MIRO and MURAL. Because of its design we opted for MURAL, and we have been able to apply it successfully. Especially in the bachelor and master programme at our university we were able to experiment and develop our skills. MURAL can be used for sticky-note sessions just like in physical settings. This needs to be facilitated by means of pre-designed templates with recognisable instructions and spaces where the virtual sticky notes can be located. While participants are able to see each other's cursor with their names on the screen, MS-Teams can be used to talk with each other at the same time. It should be noted that latterly, MURAL added a live chat function too. Results are stored in the cloud and can be exported in the form of, for example, a pdf file. MURAL offers a lot of sophisticated options, such as dot voting. To get used to the programme it is recommended to invite participants first to an exercise, that allows them to play and practice their basic skills on a specially designed template. Since the pandemic has passed, the ETFI has returned to physical settings, because they allow for much more interaction between the participants. In some cases, a combination of physical and online sessions has proven successful (hybrid workshops).

As the experiences during the COVID-19 pandemic illustrate, the ETFI's learning process is still ongoing, in an attempt to support the tourism industry with becoming more future-proof, which was the key objective when the ETFI was established. We realise that, quoting Godet and Durance, 'the quality and relevance of the scenarios is directly proportional to the knowledge and experience of those who create them. One can learn to recite by heart the methods in a few weeks but years of practice and research are needed to become a seasoned professional' (Godet & Durance, 2011: xix). We are happy that in scenario planning and strategic foresight work performed at the ETFI, we can rely on a staff team with broad knowledge and experience with the industry.

6.8 Discussion Questions

(1) Do you monitor macro-developments impacting upon tourism and how they evolve on a regular basis? If so, what are important sources for you and why?
(2) Could you identify a few blind spots that you possibly have? Why are these beyond your observation?
(3) Have you ever been surprised by unexpected events, innovations, or other wild cards? Which wildcards, and why?
(4) How far do you look into the future if you are developing plans for your business, organisation or destination, and why?
(5) Which advantages would it have to extend your time horizon with about 5–10 years?
(6) How would you incorporate this into your planning or planning cycle in a feasible way?

7 A Scenario Framework for the Post COVID-19 Futures of Tourism

Learning Points

- Scenarios for the future of tourism in a post-COVID-19 world show that there are multiple ways to recover from the impacts, other than business as usual.
- Scenarios for the post-COVID-19 future of tourism can be used to discuss the desirability of recovery and bouncing back to business as usual versus the desire to bounce forward to other scenarios.
- The study was conducted at the beginning of the worldwide COVID-19 outbreak, spring 2020, but has retained its value throughout the COVID-19 crisis.
- Scenario planning proves to be a helpful tool in crisis situations, for stakeholders in tourism to understand possible impacts, plausible futures and possible routes for recovery.

7.1 Background

The purpose of the project was to understand the possible impacts of the COVID-19 outbreak on tourism, as well as sketch a framework of four scenarios on the future of tourism in the post-COVID era. Scenario planning was selected, in particular the use of explorative scenarios, because of the unprecedented situation, disrupting many trends and causing a feeling of uncertainty. It was first and foremost selected as a sense-making device, however, the eventual outcome of the project, the framework of four scenarios, proved to create a helpful and powerful overview. The framework was intentionally created to spark discussion about possible futures, plausible futures, futures to achieve and future to avoid. Also, the framework was created with the intention to remain valid for quite some time, throughout the COVID-19 crisis, as well as to use the findings in various contexts, such as destination development, strategy building by organisation, for planning and policymaking.

The tourism industry had experienced a severe crisis due to the outbreak of the corona 'COVID-19' virus. Initially starting in December 2019 in Wuhan, China, it quickly became an unparalleled global pandemic, as a result of the exponential spread of the virus. To limit further spread of the virus many countries have proclaimed a so-called lock down. Social life has been blocked with the result that the economy has come to a grinding halt worldwide, with the tourism industry in its wake. Events have been cancelled, travelling on shorter (domestic) and longer distances (international) has become impossible and people are no longer allowed to come together in groups. As a result, public transport is no longer possible, aircraft remain on the ground, turnover of firms in tourism and leisure as well as the hospitality industry has completely evaporated, with a prospect of bankruptcies and growing unemployment.

For this project, the initial question was: how long will the virus remain among us and how long will the crisis last? This situation creates fear and uncertainty, especially because there is no clear future perspective. In such cases, it is valuable to develop explorative scenarios. Explorative scenarios outline future perspectives of the force field that a business, organisation or sector could encounter in the medium or longer term. Such scenarios do not only help to understand how the force field might develop, they also provide those involved with inspiration to think ahead, break new ground and to increase resilience and contribute to future proofing the tourism industry. Over the years, some factors may have become more important, others less. For some forces it will be fairly certain how they will develop in the coming years, for others it is extremely difficult to predict.

The project was stated because the research team at the European Tourism Futures Institute (ETFI) felt that the disruptive power of the COVID-19 crisis could be so big that it could turn out to become a game changer for many factors on which the national and international tourism industry depends. In line with Dator (2009), his understanding of a decline and collapse scenario, the tourism industry will have to deal with the potential disruptive power of COVID-19 and related uncertainties regarding the futures of tourism over the next couple of years. Explorative scenarios can be used to explore that force field and to map out the circumstances that the tourism industry could face.

7.2 Approach

The approach consisted of the consultation of a group of experts and followed the logic of the 'ETFI-scenario steps' (see Figure 7.1) where we have shown to create a set of robust scenarios that help out a range of stakeholders in tourism practice.

The group of experts were from the universities associated with the Dutch Center of Expertise for Leisure, Tourism & Hospitality

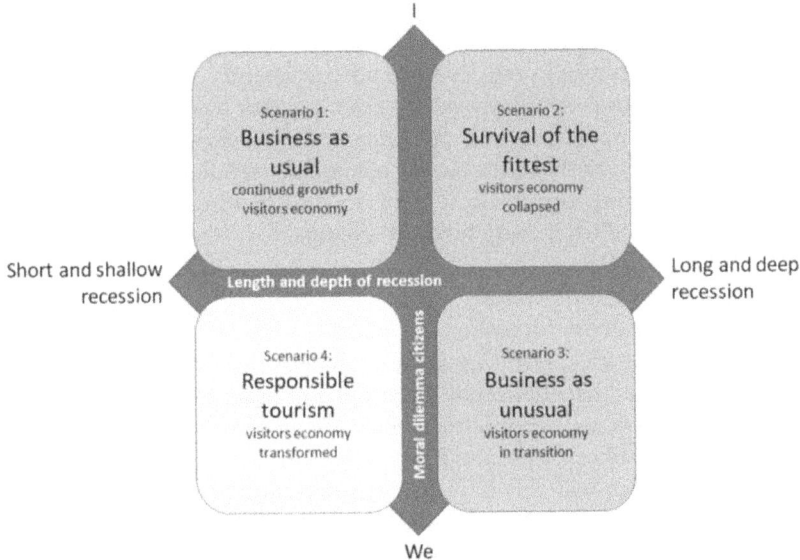

Figure 7.1 Scenario framework for post-COVID tourism

(CELTH – www.celth.nl). CELTH is a consortium in which experts collaborate in the field of leisure, tourism and hospitality from Breda University of Applied Sciences, NHL Stenden University of Applied Sciences and the associated European Tourism Futures Institute (ETFI) and HZ University of Applied Sciences and the associated Knowledge Center for Coastal Tourism (KCKT).

Above, we have mapped five key forces that we think will shape the tourism industry in the next 5–7 years. Those forces and the way in which they interact will create all kinds of uncertainties for the coming years. In order to be able to sketch a perspective, the uncertainty must first be delimited. We do this by taking the forces that we believe will have the greatest impact and at the same time are the most unpredictable, as our starting point. These forces are called the key uncertainties. There is no doubt that the extent to which the influence and unpredictability of the above-mentioned five forces may be differently perceived by governments, sector organisations or businesses in the tourism industry. Normally, in our projects we facilitate workshops in which the representatives of such stakeholders share knowledge and expertise, exchange perspectives and jointly reach a consensus. Now that the COVID-19 crisis does not allow for such workshops, as experts, we have engaged in online discussions to identify the two most powerful and unpredictable forces of importance for the development of the tourism industry in the medium term: the length and depth of the crisis and the moral dilemma of the citizen/consumer.

We have been following the media closely for several weeks. The developments and possible consequences brought up by experts in the media have been monitored closely and consistently and were linked to our own expertise. An analysis of these insights has led to a shortlist of factors that we believe will influence and determine the future of the tourism industry, without the sector being able to influence those factors itself. By mapping the relation between the factors, we were able to identify five so-called driving forces of change.

(1) **The role of nation states**
Will countries opt for reshoring, that is, withdrawing production from distant countries, resulting in de-globalisation? Will the question of geographical distribution be questioned in order to certify security of supply? How will geopolitical tensions within the European Union (EU) and between West and East on the world stage develop, and what consequences will this have for free movement of persons, open or closed borders, border controls? Will there be more solidarity or has the crisis led to more lasting tensions?

(2) **The role of the public and semi-public sector**
How will a financial buffer for the sector be developed? By general taxes or by additional tourist taxes with which funds can be filled by the market? Or by means of long-term financial injections? Will the costs be passed on to all citizens or only to the wealthiest? How will the influence of governments develop? Will they get smaller (deregulation, privatisation, market thinking, small public sector, little concern for citizens) or bigger (strong collective sector, such as in health and education, focus on public welfare, welfare state)? And, in relation to this, how will the relationship between national, provincial and local governments develop? Will governments consider man to be superior, equal or subordinate to the economy and how will this be expressed? How will cooperation and solidarity develop between governments, employer associations, trades unions, trade organisations and entrepreneurs? Will there be more solidarity or more polarisation and hostility?

(3) **The role of large (multinational) businesses and corporations**
Will they dominate the market? Will there be acquisitions and development of conglomerations? Will they remain in the mode of ceding profits to shareholders? Will they settle for less profit in times of adversity, or will they hold hands with the governments? How will investor confidence develop?

(4) **The attitude and role of the citizen since have tasted a 'different' world**
Will a new consciousness emerge? Will people develop a different perception of the feasibility of the world, and of the environment and climate? Will the way people live together change from the point of

view of social hygiene? Will more involvement, unity and solidarity arise? Will local connectedness at street, neighbourhood, district, city and/or regional levels increase and will new regional identities emerge? Will the citizen become more creative, and will he increase his ability to improvise? Will the function of social media versus traditional media change? Will the citizen make different choices as a consumer? Will he put his focus on individual and especially material prosperity or on public welfare? What role will they allow the government in relation to privacy? How will the gap and the tension between generations develop?

(5) **Length and depth of the crisis**
When will a vaccine be widely available and will social distancing be released? How will governments' debts develop and what consequences will it have? Will there be inflation or deflation, and to what extent? How will the level of prosperity and consumer confidence develop? How will the macroeconomic situation in major tourism generating countries?

It is clear that each of the aforementioned forces raise numerous questions. Questions that give an indication of the uncertainty surrounding the future.

These two key uncertainties frame, as it were, the context for four future scenarios of the tourism industry. The table below delimit the range of uncertainty of both key uncertainties. That bandwidth is determined by the plausible extremes in which those two forces could develop over a period of 5–7 years.

Key uncertainties

(1) Length and depth of the crisis.
(2) Moral dilemma of the citizen/consumer.

Key uncertainty 1: Length and depth of the crisis

Short/shallow recession	Long/deep recession
Vaccine will come on the market in 2020; lockdowns and 1.5m society come to an end; the economy will recover in 2021.	Vaccine will only become available to the world population in 2022; the COVID-19 virus returns annually in waves as seasonal flu; governments frequently call for a new lockdown; the global economy remains under pressure, and recovery will only be achieved after 2025.
– Consumer confidence restored, high level of welfare, but low level of wellbeing.	
– Elections/political shift: economy superior to man, culture and nature; neoliberalism and populism have increased further.	– Consumer confidence low, low level of welfare, but high level of wellbeing.
– Small role of governments (deregulation, market thinking, privatisation and, therefore, small/marginal public sector), care for citizens is limited.	– Elections/political shift: people, culture and nature superior to economy; socialist ideas are leading.

(Continued)

- No or limited cooperation between central governments, provinces, municipalities, employer associations, trades unions, trade organisations and entrepreneurs. - Governments and individuals create financial buffers to avoid future crises. - Public debt has increased, but political measures to weaken the debt put pressure on citizens; profits of large companies flow to shareholders. - The financial legacy of the crisis is passed on to the citizens, each gets an equal share, which reinforces the dichotomy in society between rich and poor. - Large geopolitical tensions within the EU and between world countries because countries want to protect their own economy and health; little mutual trust and solidarity; open borders and free movement are limited; production from distant countries has been withdrawn in order to guarantee security of supply (de-globalisation); more armed and cyber conflicts; US world leadership further eroded; influence of China, Russia, North Korea has increased further. - Trade and transport are flourishing; oil prices are high.	- Strong role of governments; strong public sector, financed with tax money; new type of welfare state for the protection of citizens. - Solidarity between central governments, provinces, municipalities, employer associations, trades unions, trade organisations and entrepreneurs for the common goal: joint care for society and the environment in which we live (including greening). - The parties involved share the financial resources. - Public debt has become immeasurably large; large companies, if the situation allows, donate part of their profit to the public interest, and vice versa, losses are compensated for in bad times with public money. - The financial legacy of the crisis is passed on to the citizens, proportionate to their financial means. - Geopolitical solidarity, consensus and cooperation between countries in the EU and between world countries; open borders and free movement of persons; less armed and cyber conflicts. - Trade and transport in a deep trough, which results in low oil prices.

Key uncertainty 2: Moral dilemma of citizens/consumers

I	We
- Man has not learned from the crisis and has fallen into old patterns. - Values underlying views on nature and the environment remain the same. Nature is feasible, humans are the dominant species. - The aim is individual prosperity. - Focus on material needs. Unlimited drive to consume. - People want to keep control themselves because of privacy. - Self-interest is paramount. Rebelling against the other and polarisation (race, ethnicity, gender, social class, generations, for example. GenZ vs. GenY & Z, youth versus elderly). - Social media as outlet for discontent about the other. Public sector and authorities have no respect (government, science, police, teachers, etc.).	- The crisis has brought people to repentance. - Citizens more aware of the inseparable relationship between man and nature and its effects on health. Rethinking of the value of nature, the environment, local residents, traditional media. - The aim is collective wellbeing. - Focus on quality of life. Consumption is attuned to this. - Man is willing to hand over part control to governments, despite the use of technological tools (drones, facial recognition, apps). - Common interest is paramount. 'Social hygiene': people consciously think about the implications of their own actors for others. Commitment, togetherness, solidarity, regardless of race, ethnicity, social class, generation, age. Local solidarity in street, neighbourhood, district, city and region. - Social media are social again, facilitating connections genuine between people. Public sector and authorities enjoy full respect (government, science, police, teachers, etc.).

Both key uncertainties with their delimited bandwidth can be combined in a cross. Thus, they provide a framework for four scenarios:

(1) A future scenario framed by a short/shallow crisis and an I-oriented citizen.
(2) A future scenario framed by a long/deep crisis and an I-oriented citizen.
(3) A future scenario framed by a short/shallow crisis and a we-oriented citizen.
(4) A future scenario framed by a long/deep crisis and a we-oriented citizen.

The framework shows four different contexts, each leading to a very different future perspective of the tourism industry. A suitable title has been devised for each of these scenarios. In the following picture, the two core uncertainties with their bandwidths form a cross in which the four scenarios have been positioned. Per scenario, a storyline has been created to paint a picture of the future of tourism in that particular context. Following the strategy map by Yeoman and McMahon-Beattie (2014), each scenario is systematically elaborated in terms of its guiding principle (what is leading tourism development), the visitor (what is the type of profile of the future tourist), the market (how will the demand side develop), key issues (possible issues to expect and to overcome), strategy (possible response measures), risks (possible and relatively unexpected changes that need to be taken into account for this scenario) and values (the values which are generally associated with this scenarios and are dominant among stakeholders).

7.2.1 Scenario 1: Business as usual – continued growth of the tourism industry

As soon as the crisis ends, the tourist lapses into his old behaviours. The demand for travel has accumulated into a reservoir that now suddenly 'empties'. Businesses smell their opportunities, fully respond to the reborn demand and flourish like never before. Because the recovery period has ended, the focus is again on further economic growth. Many companies are taken over by large international chains (conglomeration), but there is also room for new niches. Both travellers and businesses feel unrestrained in their behaviour. All this causes an overstrained tourism industry, heavier ecological pressure and negative social impacts. Both the positive and the negative consequences of travelling continue unabated. The mutual distrust of, and fear between, countries within the EU and beyond has sharpened relations. This has led to the reintroduction of border controls in the EU and stricter border controls outside the EU.

- **Guiding principle:** focus on back-to-business and growth, just as before and as soon as possible.
- **Visitor:** nearly the same type of visitors as before, with the same behaviour; the same hotspots overloaded; more and more often

travelling; a lot of air traffic; unlimited wanderlust; delays due to travel restrictions.
- **Market:** same markets of origin as before; in terms of size markets have quickly stabilised since the crisis; many businesses, especially family businesses, are unscathed by the crisis through a combination of their own creativity and help from governments; ongoing globalisation; large international companies and corporations dominate the market; the complexity of the large international corporations makes them vulnerable to new disruptions, profits go to shareholders.
- **Key issues:** focus on numbers; mass tourism; discussions about tourism and social impacts flare up again, as well as discussions about the increasing negative ecological impact; possibility of a return of COVID-19 has impact (1.5-metre economy); border controls; employment in the sector is growing further; distributing employment growth appropriately in time and space; more businesses work with flexible shell of workers to quickly get rid of them in case of new outbreaks (to limit fixed costs).
- **Strategy:** strong commitment to a recovery offensive; growth and profit maximisation; crowd management and crowd control; strong focus on mass and security; border control.
- **Risks:** too few lessons learned; financially the sector is still vulnerable and non-robust, due to little (financial) buffer capacity; recovery offensive too successful; stressed tourism industry; over-tourism hot topic again; scaling up of companies have made them 'too big to fail' and very costly to let them survive in times of crises (cf. KLM-Air France, Booking.com).
- **Values:** The values associated with man who dominates over nature remain intact (nature only has value if it can be used by man, nature is feasible, nature is at the service of man); the human–animal relationship is maintained; wildlife trade and exploitation of animals for profit in tourism remains commonplace, despite recurring discussions of ethical and moral dilemmas.

7.2.2 Scenario 2: Survival of the fittest – tourism industry collapsed

Citizens continue to hold on to their 'right' to vacation to faraway places, which means that the need for travel remains strong. However, the economic recession makes it financially impossible for most people to meet that need. Because people are not able to visit far off places (financial and national borders), they are looking for alternatives in and around their home/home country. The sector remains rigid in its approach. The battle for the reduced number of holidaymakers is expressed in fierce competition. Many airlines (especially low-cost airlines), tourism related businesses and catering businesses have gone bankrupt. This even goes for vital companies, because they have constantly put their money into new investments. A few large investors and players dominate the scarce market. In order to prevent over-restructuring in the sector, to prevent fragmentation and still maintain a minimal supply for its own citizens, governments have nationalised important vital players, such as national airline companies, national railway companies, but also a number of hotel, bungalow and camping chains have been taken over by the national governments (state-owned companies: in the past, certain holiday parks belonged to unions).

- **Guiding principle**: survival strategy; take what you can get.
- **Visitor**: a large share of nearby markets; market split between rich and poor, and haves and have nots. Emergence of permanent guests.
- **Market**: same markets of origin as before, however, considerably smaller in size; low-frequent travel; tourism is a luxury good; financial losses are compensated with public money (socialised costs); emerging local leisure and hospitality concepts replace former ones.
- **Key issues**: costs of the recovery operation are substantial and are recovered from the citizen, proportionate to their financial capacity; markets have become smaller; acquisitions of large players by governments; large unemployment in the sector causes additional uncertainty among entrepreneurs; mistrust among entrepreneurs regarding cooperation with other entrepreneurs (competition, no con-colleagues).
- **Strategy**: profit maximisation; competition for the scarce visitor; consumer engagement is crucial for customer loyalty.
- **Risks**: competing on price instead of quality, which has made the sector very vulnerable; emergence of a few big winners (wealthy, large players) versus many losers; major restructuring and reorganisation; organisational fragmentation, due to mutual distrust.
- **Values**: nature is used for tourism and threatened by overexploitation; in tourism animal welfare is low on the agenda, making money is the most important thing; nature's primary value is to make money from; wildlife trade continues despite threats to public health.

7.2.3 Scenario 3: Business as unusual – tourism industry in transition

The long recession has forced the holidaymaker to meet his holiday need in a different way. Governments, companies, knowledge institutions and citizens (quadruple helix) have joined forces to help to meet this need as much as possible. The parties all contribute with knowledge, subsidies, expertise and manpower. 'Under pressure everything becomes liquid' is a saying that also makes its mark here. Creativity flourishes and numerous innovations ensure a total revolution in tourism, both in terms of products and services and in terms of revenue, exploitation and management models. It represents a fundamental break from the past. It is the era of high-tech tourism that is accessible to anyone who needs it. The tourism industry gives rise to the local and regional value and production chains with legal entities, such as the cooperative as a 'renewed' exploitation model, providing purpose for society and a focus on circular production.

- **Guiding principle**: the progress of a fundamental transition.
- **Visitor**: new type of tourist and guest; purpose is central to the guest; digital; mixed reality; conscious 'quality tourist'; relatively few in number; relatively well-spent; more aware of the relationship with nature, renewed values which is expressed in, for example, more attention to animal welfare; relatively low in an ecological footprint.
- **Market**: new forms of tourism (e.g. eSports, VR and MR tourism); meaningful hospitality offering with local and regional circular products; large share of domestic tourism and day recreation, as a result of conscious and local consumption; relatively small in size; new business types (also VR); dynamic 'young!' sector.
- **Key issues**: recovery costs are substantial and are recovered from the community (or: are raised jointly by parties); companies need to go through rigorous changes, not just adapt but change thoroughly; traditional companies miss the boat and disappear from supply (bankruptcies, shifts to other sectors); governments take a greater role in exploitation.
- **Strategy**: quadruple helix; creativity and innovation; new products, services and revenue models; value-driven instead of profit maximisation; commitment to tourism in line with authenticity/sense-of-place/DNA for high-quality and credible offerings.
- **Risks**: laggards who cannot keep up with development (read: generations before GenY); a lot of temporality/high pop-up character of activities.
- **Values**: attitudes towards nature and others have changed; stakeholders in tourism realise that living respectfully with each other and with other life on Earth is important for the health of the entire planet; a transition to regenerative tourism has started, aimed at achieving positive changes in society, rather than minimising the negative effects of tourism; damage to nature is reduced, nature is given more rest in certain places and moments to recover.

7.2.4 Scenario 4: Responsible tourism – tourism industry transformed

However short, the recession has opened the eyes of the tourist. Realisation has grown that globalisation and the international travel that is annexed to it has largely contributed to the spread of COVID-19 and to the recession. The holidaymaker has become more aware of the consequences of his travels and is taking more conscious and responsible choices, based on transcending values. There is a need for purposeful products and services. Consumers are choosing purposefully and destinations that are close by, dampening the unrestrained growth of pre-crisis international tourism. To make safe and responsible choices, he relies on reliable (scientific) information. The traveller has no problem with the fact that this information has been obtained through careful monitoring of his behaviour, among other things, knowing that this benefits the traveller, the community and the destination.

- **Guiding principle**: rapid adaptation/quick fix, responsible tourism (reconsideration, discipline).
- **Visitor**: a conscious intrinsically motivated consumer; well informed; sustainable; aware of (local) impact; looking for synergy; aversion to chains/TCNs, while attracted to the local; relatively many in number; more staycations because holiday makers have increased awareness of air pollution and realise that nature close to home gives a sense of belonging, has a positive effect on people's wellbeing and stimulates social contacts and cohesion. As a consequence, their own living environment and own country as a holiday destination are revisited.
- **Market**: proximity tourism, mainly domestic tourism and day recreation; relatively large in size; air traffic has recovered somewhat, but kerosene taxes and higher taxes are putting pressure on demand; weaker companies have gone bankrupt; wealthy companies are adapting products to new market needs and are experiencing growth because of a lack of competition; regional and local pearls dominate the market.
- **Main issues**: focus on quality; vulnerability (resilience) of the tourism industry; employees remain loyal to the sector; deglobalisation to ensure supply; implementation of sustainability in companies and mobility; closing of local/regional chains; circular economy; paying more attention to human and animal welfare in tourist destinations and reducing negative effects.
- **Strategy**: rapid adaptation of companies; monitoring; clear choices by entrepreneurs: they implement concepts 100% without concessions; stimulating repeat visits by offering optimal experiences.
- **Risks**: negative consequences for society and the environment; privacy issues (big brother); demand exceeds supply, which results in a price increase of the best and most popular places.
- **Values**: growing awareness that man is part of nature, and nature is part of man and that not everything is controllable; that living together with each other and with other life on earth with more respect is important for the health of the entire planet. The virus has taught people that we are a community (social capital).

7.3 Outcomes of the Study

Each out of the four scenarios provides a plausible picture of what the tourism industry might look like in 2025. Please note: what it *might* look like. As noted in the introduction, the scenarios presented are neither predictive, nor goal-based, but explorative scenarios. Because they are not predictive scenarios they are not designed to state or predict what will happen. Because they are not goal-based scenarios, we cannot choose one and derive a strategy or action plan from it to shape the future. Instead, the scenarios are explorative scenarios. Each scenario explores a different probable, plausible future perspective of the tourism industry.

The uncertainties that form the basis for the scenarios are subjective. Different parties – entrepreneurs, governments, financiers, DMOs – may see things differently. What is predictable for one is unimaginable for the other. What seems as clear as a lump to one person is a great uncertainty to another. That is why a conversation between those involved is important. To share knowledge and insights but also to learn from each other and discover that there are other ways to look at the world around us. After all, each of us has developed a dominant way of thinking and solving problems, which sometimes prevent us from looking at our environment with different eyes.

The scenarios are also a snapshot. Although they take future developments into account, our view of what may happen in the future may change from time to time. Because our insights change, because new developments are imminent, because a major event may occur, because an important new innovation has been brought about, because the next disruptive force is emerging, because… you can fill this in yourself. It is, therefore, important to continuously monitor and try to understand the early signals of new developments in the complex force field impacting upon the tourism industry. Scenarios can then be regularly calibrated to those new expectations and possibly adjusted. The adapted scenarios can be used to test the existing strategies and measures for robustness and resilience, but they can, therefore, also be used to change or renew the policies and strategies. This requires adaptive capacity and contributes to a greater resilience of developments that may come our way in the future.

7.4 Concluding Remarks

When we see the scenarios as probable, plausible futures, we realise that they provide insight into the possible playing field of tomorrow and that they can help us to prepare for such a playing field. Moreover, when we understand that the future may contain elements of each of the scenarios, together they can inspire us to develop strategies and actions to prepare us for all possible futures.

7.4.1 Organisation: Mobilise your network and take the following steps

In this study we have outlined four scenarios for the future of the tourism industry in 2025. We see this as a first step in talking with representatives and stakeholders of the tourism industry – from governments, businesses, trade organisations, financial institutions to DMOs – about the value of these scenarios and how to develop them into usable future perspectives that extends beyond the short-term solutions that dominate the media today.

7.4.2 Inspiration: Immerse yourself in the forcefield and the scenarios and let them inspire you

It would be good to delve into the force field and into the scenarios. How do you position yourself? Does it fit your view of what could happen next? Do you consider the scenarios plausible? Why yes or why not? You could engage with colleagues and other stakeholders to compare and discuss your perspective with the views of others. You may come to the conclusion that the scenarios can be tightened or otherwise adjusted in certain areas. If so, we would like to hear from you.

7.4.3 Tooling: Apply the forcefield and scenarios to your business, your organisation, your sector or destination and identify the possible consequences

While the scenarios are not hard-nailed predictions and are based on subjective assessments of the world around us, they can be perfectly used to review your current business or organisation, your policy, sector or destination. What fits, what does not fit? Why? What are the possible consequences of this? What can, or should, you do to be able to anticipate such developments? Would that be feasible? What risks do you run if you do or don't do that?

7.4.4 (Re)orientation: Use the scenarios as a source of inspiration for creative and innovative renewal

You can use the scenarios to use as a source of inspiration to creatively and innovatively reinvent your business or organisation, your sector or destination, to review your policy, or to look for support measures. It may result in a new concept, a new revenue model, new products or services or other strategic choices that fit the requirements of the new situation. Here, too, you can ask yourself whether such an idea fits into the future context outlined in the scenarios; whether it fits the goals of your organisation/company and is accepted by shareholders and other stakeholders; whether it meets the needs of the future market; how sustainable it is in view of the economy, society and the environment; whether the idea is feasible, not only financially but also in terms of, for example, manpower and competences; what risks you may run; and whether your idea can be quickly scaled up or down when market conditions or beyond force you to do so.

7.5 Discussion Questions

(1) The scenarios that are presented in this chapter are all plausible tourism futures. In practice, individuals might consider one more plausible than others, more realistic, more desirable. How does this look in practice? How can you use a scenario framework (a 2 × 2 matrix, as presented in Figure 7.1) to broaden an individual's perspective and let them see the opportunities encapsulated in the other, less conservative, less comfortable tourism futures?
(2) How can you tailor or specify a somewhat generic set of future scenarios for tourism to specific subsectors?
(3) How long are scenario frameworks (2 × 2 matrices) valid? Until what changes in society do they lose their relevance?
(4) If not even a global crisis such as the COVID-19 pandemic can shift the tourism industry to adopt a fundamentally different development model, more in line with scenario 3 and 4 in Figure 7.1, what will? And how can scenario planning play a role?

8 The Tourism Futures of Rural Friesland: An Integrated Spatial Planning Approach to Tourism Planning

Learning Points

- The tourism futures of rural Friesland are dependent on choices made in other (policy) domains that are, in turn, contingent on how forces of change play out.
- The main important uncertainties for rural Friesland are scale, upscaling versus downscaling and how we manage the impacts of climate change, by means of adaptation or mitigation.
- The tourism futures for rural Friesland show possible futures but also trigger discussion on possible futures to achieve and futures to avoid.
- The scenario planning approach results in a 'mixing console' that help (public sector) stakeholders to apply to their locally specific situation.

8.1 Background

The case study on the tourism future of rural Friesland presents a set of future scenarios for the province of Friesland, the Netherlands. The scenarios address regional development in general and are, therefore, highly relevant to tourism development and planning. The scenarios show huge implications for the future of tourism and highlight the importance of taking a multi-domain, integral approach to the development of tourism and regions at large. The scenarios were created as part of a project in the context of a wider program called the 'Omgevingslab Fryslan'. Within this program, a range of dialogs were organised about important topics for the 'Omgevingsvisie' (vision on the environment), a mandatory requirement by the Dutch National law on spatial planning. The vision includes a perspective on how the future of the province is going to look, according to the provincial authority. Once the vision is formally approved by the provincial council of publicly elected representatives, an ordinance follows that serves as the legal basis.

The 'Omgevingslab Fryslân' was initiated by a group of public sector stakeholders, collectively known as 'De Friese Aanpak' (The Frisian Approach), combining all the municipalities, the provincial authority and the water board (Wetterskip Fryslân). Each of these public sector stakeholders had to make an 'Omgevingsvisie'. The program 'Omgevingslab Fryslan' was initiated to set up dialogs with the community (in the Frisian local language known as 'Mienskip') about the future of the province of Friesland. Central to the dialogs were topics, such as the energy transition, about cities and villages and about government and governance among others. In the dialogs, various challenges were addressed. What trends and developments are taking place and will take place which we need to be taken into account? What are the consequences for the province of Friesland? What are the opportunities? What are the challenges? What is desirable, what is undesirable? The output of the dialogs was considered as building blocks for the various public-sector stakeholders, to help design their individual 'Omgevingsvisie'. The website https://defrieseaanpak.nl/resultaten-labs/omgevingslab-fries-platteland served as a digital platform for bringing together and disseminating the information. Within the program 'Omgevingslab Fryslan' (later repositioned as 'De Friese Aanpak' – which translates to English as the 'The Frisian Approach'). The team of researchers of NHL Stenden, its European Tourism Futures Institute (ETFI), were responsible for the project 'Omgevingslab Fries Platteland' – freely translated as the dialogs about the future of rural Friesland.

8.2 Approach

The approach was part of a complex multi-actor process, embedded in a wider policymaking project ('Omgevingslab Fryslan'). For the future scenarios, the ETFI organised the steps taken in the workshops following the logic of the 'ETFI-scenario steps' (see Figure 8.1).

Scenario planning was central to the ETFI project 'Omgevingslab Fries Platteland'. Scenario planning was selected, since society is dynamic and complex. Moreover, Friesland will see the impacts of developments in terms of climate, demography and technology. These developments take place more or less autonomously and can be forces driving change at local or regional levels. It is possible to explore such development and anticipate which might be coming. The ETFI's scenario cycle was used to structure the scenario planning approach.

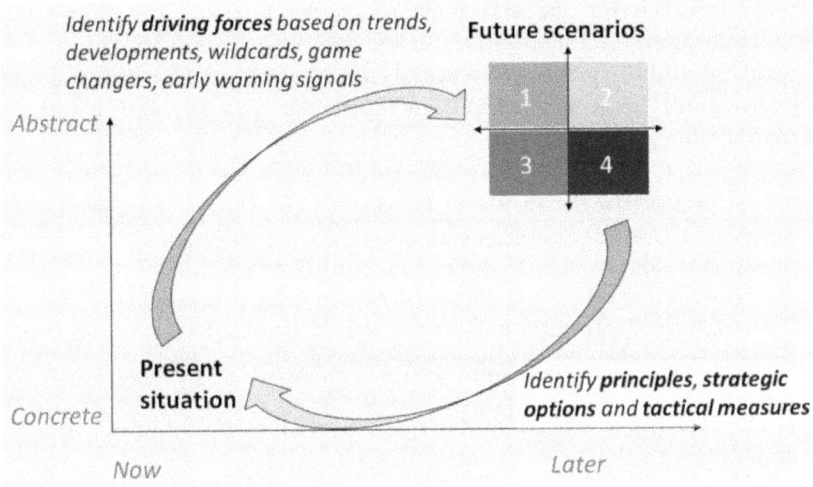

Figure 8.1 Scenario planning process

We organised five workshops with a variety of stakeholders to identify trends and development. The input was used to design a scenario framework. Subsequently, another five workshops were organised to identify the implications and policy actions. Central to the process was the scenario question: 'how can we ensure that in the coming 15 years all necessary functions of rural Friesland go together with the qualities of the environment?' Below the figure, a detailed description of the steps is presented.

8.2.1 Step 1: Looking ahead to the future

In June 2017, the first round of in total five regional workshops were organised. The workshops were held in various part of the province of Friesland but were structured identically (see below). Participants were challenged to look at Friesland from the outside in, to identify the trends and developments that take place in society, and that could potentially affect the spatial and regional development of the province. The central, guiding question we posed to the participants was: *which trends and development will influence, from the outside in, the functions of rural Friesland?* The guiding question was prepared in a meeting on 1 May 1 2017, by the support committee of our project 'Omgevingslab Fries Platteland'. The committee consisted of one representative of the provincial authority, one for the Frisian municipalities, one for the water board (Wetterskip Fryslân) and the principal investigators of NHL Stenden, its European Tourism Futures Institute (ETFI). The guiding question was further discussed by a wider, sounding-board group, installed for the sole purpose of the project, consisting of a wider group of representatives from the Frisian municipalities, the Provincial authority, the water board, the Netherlands Agricultural and Horticultural Association (LTO) and the Federation for Nature and the Environment (NMF).

Workhops 'Omgevingslab Fries Platteland' – Round 1

- Tuesday, 20 June 2017 – region of the Wadden Sea Islands (location: Leeuwarden)
- Tuesday, 20 June 2017 – region of southeast Friesland (location: Drachten)
- Wednesday, 21 June 2017 – region of northeast Friesland (location: Dokkum)
- Thursday, 22 June 2017 – region of southwest Friesland (location: Sneek)
- Thursday, 22 June 2017 – region of northwest Friesland (location: Franeker)

The structured procedure during each workshop of round 1 was as follows:

(1) Participants started with naming the trends and development that influenced rural Friesland from the outside in, which the Frisian stakeholders should take into account. Participants started individually. Each trend or development was written down on a separate sticky note and place on the wall. This process resulted in an enormous overview.
(2) Participants presented the results to each other and continued by clustering related results. Each cluster was given a name that fitted well with the theme of the cluster. These clusters were, from that point onwards, treated as 'forces driving change'.
(3) For each cluster, participants named the extremes in which the force could play out in the next 15 years.
(4) Participants ranked the clusters in terms of their importance (by means of voting, using green stickers) and degree of relative uncertainty (by means of voting, using red stickers).

Via this structured procedure, participants were able to identify the most important and uncertain forces that drive change, which we call the 'key uncertainties'. Having held five workshops, we gathered a huge number of trends and developments, as well as numerous forces driving change. Nevertheless, there was some overlap and somewhat of a discourse could have been identified within the workshops results. In a working session by the support committee, the results were intensively discussed so as to filter out two key uncertainties that could be used to construct a scenario framework. The two key uncertainties became 'scale', accompanied by the extreme outcomes of large scale versus small scale, and 'dealing with consequences of climate change', accompanied by the extreme outcomes of climate adaption versus climate mitigation.

The Tourism Futures of Rural Friesland 103

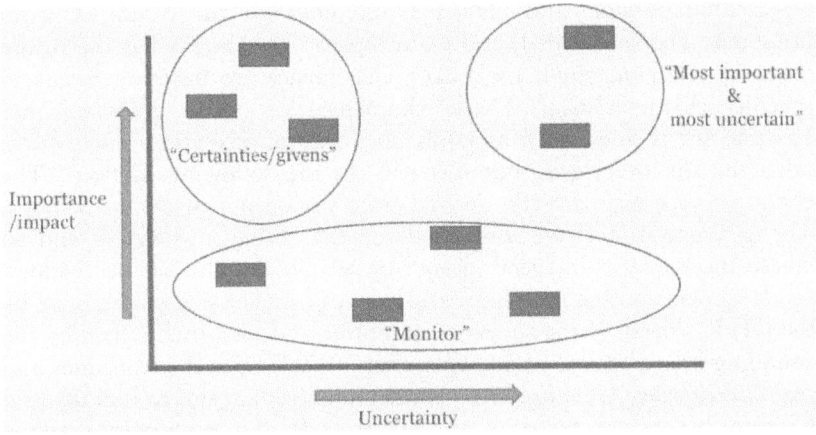

Figure 8.2 Importance by uncertainty matrix

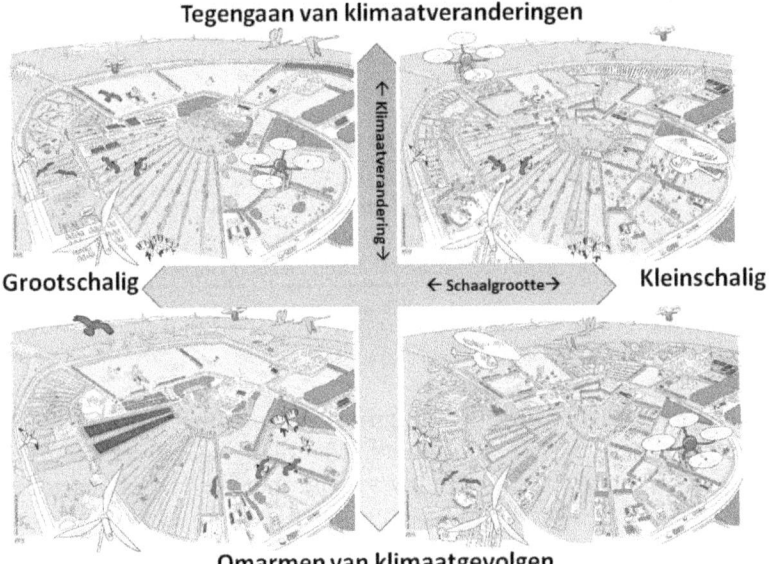

Figure 8.3 Scenario framework
Source: Studio Ronald van der Heide.

The key uncertainties were used to create the scenario framework, as presented above. Other forces driving change, or other trends and developments, were used to further elaborate the individual scenarios.

8.2.2 Step 2: Future scenarios

The key uncertainties and accompanying possible direction for development were combined in a scenario framework. The scenarios represented possible and plausible outcomes for the future of rural Friesland. The scenarios literally and figuratively show what the future of rural Friesland could look like. The images are based on scenario specific characteristics. These characteristics were described per scenario for themes, such as landscape, nature, recreation, energy, etc. (also see the overviews, per scenario, in the following section). The characteristics were directly derived from the output of the workshops. The characteristics were also used for the scenario storyline and to crease the scenario imagery (design by Studio Ronald van der Heide). Draft versions of the storylines and scenario imagery were prepared by the ETFI, refined by the support committee and, finally, enriched by the sounding-board group, on 14 September 2017. Finally, the storylines and scenario imagery were used in a short animation, again, to literally and figuratively show participants, and the wider Frisian community, how the future of rural Friesland could look.

8.2.3 Step 3: Looking back from the future to the present

In October 2017, a second round of workshops were held with the Frisian community.

Workhops 'Omgevingslab Fries Platteland' – Round 2

- Wednesday, 4 October 2017 – northeast Friesland (location: Dokkum)
- Wednesday, 4 October 2017 – region of southwest Friesland (location: IJlst)
- Thursday, 5 October 2017 – region of northwest Friesland (location: St. Annaparochie)
- Monday, 9 October 2017 – region of southeast Friesland (location: Drachten)

The structured procedure during each workshop of round 1 was as follows:

(1) Participants elaborated, per scenario, on the possible consequences for the future of rural Friesland. Via the World Café method,[1] participants had the opportunity to contribute to each scenario. At the start, three key themes were pre-given – nature, agriculture and recreation/tourism – however, participants added additional themes which they felt were needed. First, participants wrote down on sticky notes how the future that was presented in the scenario could result in change. Second, participants shared their ideas to further discuss the direct and indirect consequences of such changes.
(2) Participants identified the desirable consequences and undesirable consequences, by means of voting, using green and red stickers, respectively. The results provided valuable insights into possible opportunities, threats, action and other implications that participant associated with the scenarios.
(3) In a plenary round off, participants openly and elaborately discussed their views on the most desirable and the most realistic scenario.

8.2.4 Step 4: Analysis of the results

The project 'Omgevingslab Fries Platteland' resulted in a range of important themes and dilemmas, regarding the future of rural Friesland. During the analysis of the results, we identified 10 different themes (e.g. energy, recreation and tourism, nature, liveability, etc.) and, per theme, a set of dilemmas that relate to them, based on the forces driving change and their alternative, extreme outcomes. How the dilemmas play out, depends on how the forces that drive change evolve over time. Alternatively, how forces driving change are dealt with can determine how the future of rural Friesland will look. As such, the overview was named the 'mixing console' for rural Friesland. The mixing console consist of themes and, per theme, a set of dilemmas. For each dilemma, choices have to be made about the future. How the mixing console is configured partly depends on societal dynamics (autonomous change) and partly by decision making (induced change). The mixing console is quite comprehensive, which is a good reflection of the substantive number of topics that came up during the workshops and the (complex) number of options stakeholders should address. On the next pages, details are given about the four scenarios. For each scenario we show how the mixing console is configured – albeit in a very general way.

8.2.4.1 Scenario 1: 'We win the battle of the water and tackle it on a large scale'

Figure 8.4 Scenario 1
Source: Studio Ronald van der Heide.

Storyline

It is the year 2030. In the battle against the water, Friesland is a frontrunner. We put safety first. The sea is no danger because of the high dikes, meaning we can make effective and efficient use of each plot of land. We think big and act accordingly. This is represented by the straight lines in the landscape, full commitment to sustainable technologies, but also the clustering of important economic sectors. Farms are mega-companies that produce efficiently for the world. Large parts of natural areas are reserved for biological farms. The nation in Friesland is unique and attracts many visitors. They enjoy themselves, offer for multiple days, in the large scale and internationally renowned accommodations are the coastal zones and in-land waters. Mobility is modern, smart technology is monitoring production continuously and part of the landscape is designed for the energy transition. Friesland is almost energy neutral, due to clear, centrally organised interventions.

Characteristics of scenario 1

The characteristics are derived directly from the results from the round 1 workshops (June 2017) and were used as the basis for the storytelling and animation of this scenario.

- **Landscape**: large-scale, efficient, effective, monocultures, dry feet, landscape can be made and controlled, sand nourishments to maintain Wadden Sea Islands, salinisation is a given.
- **Coastal defence**: raising, strengthening and widening dikes 'just in case' a flood occurs. Guaranteeing safety, safety first. Sea level rise is not a concern due to safety guarantee. The resulting safe, stable situation allows long-term investments inland.
- **Water**: controlled supply/discharge of freshwater, centrally organised, water level in the IJsselmeer raised to create a search water basin.
- **Nature**: large areas, combined and clearly recognisable as nature areas. This change results in a shift in flora and fauna.
- **Agriculture**: stable situation, long-term choices possible (in terms of crop types, investments), focused on world food supply, building on past successes (dairy farming, potatoes, etc.), large companies taking the upper hand over smaller, family companies.
- **Recreation**: mass tourism, strengthening recreation clusters, resorts with indoor facilities (climate-independent and, therefore, climate-resistant offer, use of technologies, such as Augmented Reality), guaranteeing swim water quality at specific central locations, large operators with a large portfolio (managing the entire chain).
- **Centrally regulated energy production landscape**: clustered wind farms, (large) tidal power stations.
- **Technology**: expanding technology at the service of past successes, for example, through robotisation, drones, precision agriculture and

self-driving vehicles. Technology used on a large scale to combat climate change, for example, through CO_2 reduction, switching from fuel to electric and self-sufficient zero-on-the-meter buildings.
- **Governance/management**: uniform approach and central interventions, building on past successes (dairy farming, potatoes). Control and management oriented.

Configuration of the 'mixing console' for scenario 1

Nature	
1. Large, contiguous areas	1. Network structure
2. Spatially separating functions	2. Interwoven functions
3. Biodiversity is declining	3. Biodiversity is increasing
4. Dehydration	4. Wetting
5. Healthiness for flora and fauna	5. Human health
6. Value creation: usable raw material	6. Valuable
7. Park landscape ('de-wilding')	7. New wilderness ('re-wilding')
Recreation & tourism	
1. Mass	1. Niche
2. Focus, clustering	2. Scattered
3. Large scale ('big attractors')	3. Small scale
4. Permanent	4. Pop-up/Temporary
5. Virtual reality (= place independent)	5. Augmented reality (= place dependent)
6. Routes and arrangements (connecting, marketing)	6. Development of physical supply (improve, renew, add)
7. Entertainment (= place independent)	7. Area-oriented (= place dependent)
8. Water recreation	8. Recreation on land
9. International market/Orientation	9. Local-regional market/orientation
Landscape	
1. Hard coastline	1. Soft transition
2. Separating functions	2. Interwoven functions
3. Monotonous landscape	3. Varied landscape
4. Production landscape/Industrial landscape	4. Residential landscape/Experience environment
5. Static landscape	5. Dynamic landscape/Adaptive landscape
Agriculture	
1. Diversification/Multifunctional	1. Monoculture/Monofunctional
2. Global market	2. Regional chains
3. New emerging shapes	3. Preserving/Nurturing traditional forms
4. Nature-exclusive	4. Nature included
5. Regeneration: interventions	5. Degeneration: depletion of the soil

(*Continued*)

Agriculture

6. Extensive agriculture	6. Intensive farming
7. Biological-ecological	7. Industrial (GM)
8. Small scale	8. Large scale
9. Scaling down	9. Scaling up
10. Area cooperatives	10. Chain managers
11. Cottage industries	11. Vacant agricultural buildings

Climate & water

1. Based on technical solutions/processes	1. Starting from natural solutions/processes
2. Central (government)	2. Local (cooperatives, individuals)
3. Big water task	3. Small-scale solutions
4. Dispose of water/store regionally	4. Retain water/store locally
5. Preventing consequences	5. Embracing the consequences: rewetting, dehydration, salinisation
6. Fossil fuels	6. Renewable, sustainable resources

Economy

1. New emerging shapes	1. Preserving/Nurturing traditional forms
2. New professions	2. Disappearing professions
3. Specialised knowledge	3. Diversification
4. Craft	4. Automation/Robotisation
5. Large scale is vulnerable	5. Large scale is efficient
6. Small (SME, family) companies	6. Major international players
7. Scale advantage ends up with citizens	7. The advantage of scale ends up with the company
8. Chain managers	8. Area cooperatives
9. Young people	9. Elderly
10. Small number of large employers	10. Broad base of small SME's

Energy

1. Slow transition (small steps)	1. Rapid transition (big bang)
2. Dependent on other parties	2. Self-sufficient
3. Revenues to energy companies	3. Revenues to communities (individuals, cooperatives)
4. Large-scale, clustered (wind farms, solar farms)	4. Small-scale, distributed (loose windmills, each with a solar panel)
5. Fossil fuels	5. Sustainable sources: wind, water, sun

Liveability

1. Vital villages	1. Shrinking villages
2. Center facilities and housing in cores	2. Facilities and living in villages
3. Forced self-reliance	3. Selected self-reliance
4. Major transport arteries (roads, public transport)	4. Fine-mesh structure (roads, public transport)
5. New residents (migrants, international students)	5. Departing residents

Governance	
1. Central control	1. Area-oriented/Self-sufficient villages
2. Monopolies	2. Freedom of choice
3. Control	3. Laissez faire/Every man for himself
4. Control/Involvement in organisations	4. Control/Involvement with residents
5. Trust and leave over to others	5. Distrust and control
6. Knowledge and information accessible to a selective group	6. Knowledge and information accessible to everyone

8.2.4.2 Scenario 2: 'We win the battle of the water and everyone shapes it themselves'

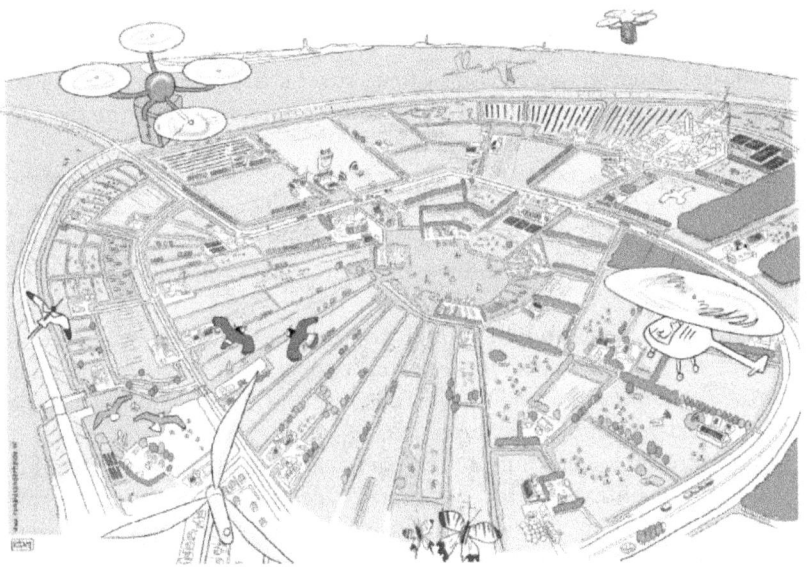

Figure 8.5 Scenario 2
Source: Studio Ronald van der Heide.

Storyline

It is the year 2030. Friesland has cleverly moved along with the changing nature. Cities and villages are protected by high dikes, while places with more space are protected by double dike lines. The Frisian countryside is very diverse and the various area types are intertwined. Due to the great variety of the landscape, we need technology that helps us make a local impact in a smart way. For example, wetting and desiccation are tackled per area. Water cooperatives are themselves responsible for maintaining the water supply at the micro level. This makes the conditions extremely suitable for organic and ecological farming. Visitors come to Friesland with varying needs. The offer is climate independent. One comes for nature and water features, the *next for indoor sports and another for innovations in agriculture, coastal defence and hydraulic engineering. Entrepreneurial individuals and organisations can indulge themselves and together they take responsibility for how the landscape develops. Friesland is almost energy neutral, and each village or group of villages implements the energy production in its own way.*

Characteristics of scenario 2

The characteristics are derived directly from the results from the round 1 workshops (June 2017) and were used as the basis for the storytelling and animation of this scenario. Landscape: varied, varied functions, small scale, interwoven functions. Agriculture, nature, recreation are all intertwined.

- **Coastal defence:** safety comes first, coastal defence is geared to the specific local situation in terms of landscape features and the risks for people, animals and companies.
- **Nature:** preserve flora and fauna in smaller, varied areas that together form an intricate and overlapping network. To counter the impact of climate change, extensive desiccation or uncontrolled waterlogging is prevented. Nature is intertwined with, and spills over into, other functions.
- **Water:** local measurement and control technology are geared to diversity of functions, local water cooperatives that can regulate arrow, quantity and quality, that can respond with micro-precision to local and very specific requirements.
- **Agriculture:** local solutions for desiccation, rehydration. There is a strong context-specific approach. Technology enables real-time monitoring of production conditions and self-intervention. There is no dependence on a water board. Climate change is counteracted, which is also reflected in a strong desire to build on past successes (dairy farming, potatoes) but due to the small scale, mainly in a bio/eco atmosphere.
- **Recreation:** climate-independent and, therefore, climate-proof offer (indoor or all-weather activities), focus on a very diverse series of niche tourism because recreation is part of, and intertwined with, everything, including agriculture, coastal defence/hydraulic engineering ('policy tourists'), nature, world heritage. Incidentally, these are not at risk from interventions in coastal defences and interventions to combat changes in landscape and nature due to climate change.
- **Energy landscape:** sustainability at the forefront with implementation at individual/company level (solar panels, zero-on-the-meter buildings).
- **Technology:** applications are adopted that serve to enable small-scale operations.

- **Governance/management:** central choices with an eye for local, specific situation. Control and management oriented.

Configuration of the 'mixing console' for scenario 2

Nature	
1. Large, contiguous areas	1. Network structure
2. Spatially separating functions	2. Interwoven functions
3. Biodiversity is declining	3. Biodiversity is increasing
4. Dehydration	4. Wetting
5. Healthiness for flora and fauna	5. Human health
6. Value creation: usable raw material	6. Valuable
7. Park landscape ('de-wilding')	7. New wilderness ('re-wilding')

Recreation & tourism	
1. Mass	1. Niche
2. Focus, clustering	2. Scattered
3. Large scale ('big attractors')	3. Small scale
4. Permanent	4. Pop-up/Temporary
5. Virtual reality (= place independent)	5. Augmented reality (= place dependent)
6. Routes and arrangements (connecting, marketing)	6. Development of physical supply (Improve, renew, add)
7. Entertainment (= place independent)	7. Area-oriented (= place dependent)
8. Water recreation	8. Recreation on land
9. International market/orientation	9. Local-regional market/orientation

Landscape	
1. Hard coastline	1. Soft transition
2. Separating functions	2. Interwoven functions
3. Monotonous landscape	3. Varied landscape
4. Production landscape/Industrial landscape	4. Residential landscape/Experience environment
5. Static landscape	5. Dynamic landscape/Adaptive landscape

Agriculture	
1. Diversification/Multifunctional	1. Monoculture/Monofunctional
2. Global market	2. Regional chains
3. New emerging shapes	3. Preserving/Nurturing traditional forms
4. Nature-exclusive	4. Nature included
5. Regeneration: interventions	5. Degeneration: depletion of the soil
6. Extensive agriculture	6. Intensive farming
7. Biological-ecological	7. Industrial (GM)

(Continued)

Agriculture

8. Small scale	8. Large scale
9. Scaling down	9. Scaling up
10. Area cooperatives	10. Chain managers
11. Cottage industries	11. Vacant agricultural buildings

Climate & water

1. Based on technical solutions/processes	1. Starting from natural solutions/processes
2. Central (government)	2. Local (cooperatives, individuals)
3. Big water task	3. Small-scale solutions
4. Dispose of water/store regionally	4. Retain water/Store locally
5. Preventing consequences	5. Embracing the consequences: rewetting, dehydration, salinisation
6. Fossil fuels	6. Renewable, sustainable resources

Economy

1. New emerging shapes	1. Preserving/Nurturing traditional forms
2. New professions	2. Disappearing professions
3. Specialised knowledge	3. Diversification
4. Craft	4. Automation/Robotisation
5. Large scale is vulnerable	5. Large scale is efficient
6. Small (SME, family) companies	6. Major international players
7. Scale advantage ends up with citizens	7. The advantage of scale ends up with the company
8. Chain managers	8. Area cooperatives
9. Young people	9. Elderly
10. Small number of large employers	10. Broad base of small SME's

Energy

1. Slow transition (small steps)	1. Rapid transition (big bang)
2. Dependent on other parties	2. Self-sufficient
3. Revenues to energy companies	3. Revenues to communities (individuals, cooperatives)
4. Large-scale, clustered (wind farms, solar farms)	4. Small-scale, distributed (loose windmills, each with a solar panel)
5. Fossil fuels	5. Sustainable sources: wind, water, sun

Liveability

1. Vital villages	1. Shrinking villages
2. Center facilities and housing in cores	2. Facilities and living in villages
3. Forced self-reliance	3. Selected self-reliance
4. Major transport arteries (roads, public transport)	4. Fine-mesh structure (roads, public transport)
5. New residents (migrants, international students)	5. Departing residents

The Tourism Futures of Rural Friesland 113

	Governance
1. Central control	1. Area-oriented/Self-sufficient villages
2. Monopolies	2. Freedom of choice
3. Control	3. Laissez faire/Every man for himself
4. Control/Involvement in organisations	4. Control/Involvement with residents
5. Trust and leave over to others	5. Distrust and control
6. Knowledge and information accessible to a selective group	6. Knowledge and information accessible to everyone

8.2.4.3 Scenario 3: 'We embrace the water and innovate with major players'

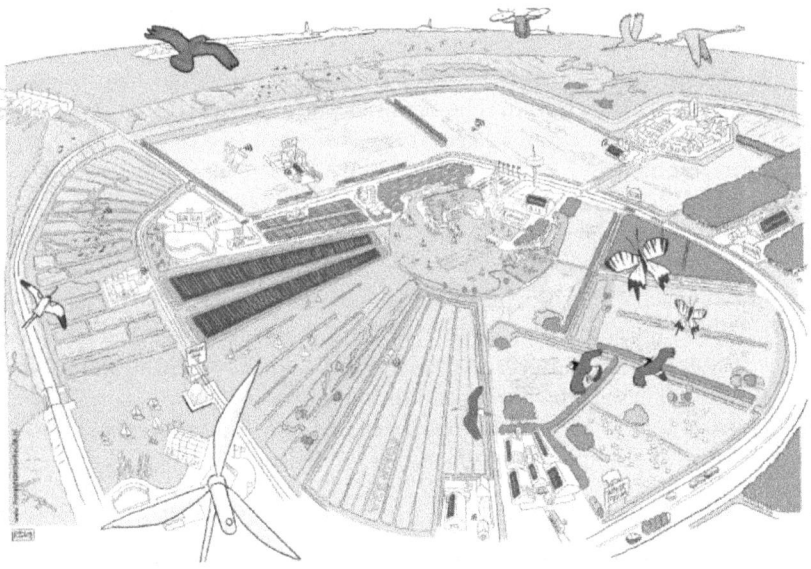

Figure 8.6 Scenario 3
Source: Studio Ronald van der Heide.

Storyline

It is the year 2030. Friesland loves the water and has made it its strength. Due to climate change, the amount of water has increased enormously. The coastline is no longer one high dike, but a soft and gradual transition from water to land. The agricultural sector grows all kinds of new crops that can tolerate the new climate through genetic engineering. A few large parties have seized their opportunities and are in control of the chain. Frisian organic food is traded all over the world. The changing climate makes it possible to use all kinds of energy *sources on a large scale. Think of solar energy, wind energy, energy from fresh-salt transitions and energy from biofuels and residual flows. Due to its scale and size, this is only possible for large companies and cooperatives with sufficient capital and leadership to control the chain. The new nature and the varied climate led to a diverse tourism industry. The rehydration of the countryside is causing a revival of the water sports sector. New concepts are tested on a large scale, rolled out in the region and exported abroad.*

Characteristics of scenario 3

The characteristics are derived directly from the results from the round 1 workshops (June 2017) and were used as the basis for the storytelling and animation of this scenario.

- **Landscape:** rehydration and desiccation are discussed at the same time. In places, the landscape becomes saline and elsewhere the soil subsides (or, both) due to the natural processes, but also through gas and salt extraction. The result is a more diverse landscape than in scenario 1, certainly wetter due to rising sea levels and declining hinterland.
- **Coastal defence:** embracing water, wider coastal zone next to high dikes, using new salt marshes for coastal defence ('building with nature').
- **Water:** accepting risks, periods of drought/wetting, sometimes water surplus and sometimes water scarcity, shortage of fresh water, soil becomes salinised, freshwater wastage is a sin, compensation fund for water damage, water storage/water retention.
- **Nature:** due to the small scale, there is more variation. This means a shift in flora and fauna. These are given space in large, contiguous nature reserves.
- **Agriculture:** climate change necessitates a general transition that includes space for saline cultivation, peat moss, bulrush and heat-resistant crops. Gene technology is applied, and industry chains are created at the level of Friesland or a mega company, land consolidation is (again) applied to make large-scale possible. Land subsidence occurs but is responded to by using lighter vehicles.
- **Recreation:** climate tourism (drought/heat, storms, rehydration), rehydration provides a revival of the water sports sector that is developing as an R&D industry for the export of products and services (in line with yacht building, kite surfing and soon hydrofoils, solar boats).
- **Energy landscape:** space for fresh/salt, biofuels (crops, algae).
- **Technology:** at the service of creating new possibilities that match the water landscape.
- **Governance/management:** support new functions, stimulate transitions in recreation, agriculture, energy, water management.

Configuration of the 'mixing console' for scenario 3

Nature

1. Large, contiguous areas	1. Network structure
2. Spatially separating functions	2. Interwoven functions
3. Biodiversity is declining	3. Biodiversity is increasing
4. Dehydration	4. Wetting
5. Healthiness for flora and fauna	5. Human health
6. Value creation: usable raw material	6. Valuable
7. Park landscape ('de-wilding')	7. New wilderness ('re-wilding')

Recreation & tourism

1. Mass	1. Niche
2. Focus, clustering	2. Scattered
3. Large scale ('big attractors')	3. Small scale
4. Permanent	4. Pop-up/Temporary
5. Virtual reality (= place independent)	5. Augmented reality (= place dependent)
6. Routes and arrangements (connecting, marketing)	6. Development of physical supply (Improve, renew, add)
7. Entertainment (= place independent)	7. Area-oriented (= place dependent)
8. Water recreation	8. Recreation on land
9. International market/orientation	9. Local-regional market/orientation

Landscape

1. Hard coastline	1. Soft transition
2. Separating functions	2. Interwoven functions
3. Monotonous landscape	3. Varied landscape
4. Production landscape/Industrial landscape	4. Residential landscape/Experience environment
5. Static landscape	5. Dynamic landscape/Adaptive landscape

Agriculture

1. Diversification/Multifunctional	1. Monoculture/Monofunctional
2. Global market	2. Regional chains
3. New emerging shapes	3. Preserving/Nurturing traditional forms
4. Nature-exclusive	4. Nature included
5. Regeneration: interventions	5. Degeneration: depletion of the soil
6. Extensive agriculture	6. Intensive farming
7. biological-ecological	7. Industrial (GM)
8. Small scale	8. Large scale
9. Scaling down	9. Scaling up
10. Area cooperatives	10. Chain managers
11. Cottage industries	11. Vacant agricultural buildings

Climate & water

1. Based on technical solutions/processes	1. Starting from natural solutions/processes
2. Central (government)	2. Local (cooperatives, individuals)
3. Big water task	3. Small-scale solutions
4. Dispose of water/Store regionally	4. Retain water/Store locally
5. Preventing consequences	5. Embracing the consequences: rewetting, dehydration, salinisation
6. Fossil fuels	6. Renewable, sustainable resources

Economy

1. New emerging shapes	1. Preserving/Nurturing traditional forms
2. New professions	2. Disappearing professions
3. Specialised knowledge	3. Diversification
4. Craft	4. Automation/Robotisation
5. Large scale is vulnerable	5. Large scale is efficient
6. Small (SME, family) companies	6. Major international players
7. Scale advantage ends up with citizens	7. The advantage of scale ends up with the company
8. Chain managers	8. Area cooperatives
9. Young people	9. Elderly
10. Small number of large employers	10. Broad base of small SME's

Energy

1. Slow transition (small steps)	1. Rapid transition (big bang)
2. Dependent on other parties	2. Self-sufficient
3. Revenues to energy companies	3. Revenues to communities (individuals, cooperatives)
4. Large-scale, clustered (wind farms, solar farms)	4. Small-scale, distributed (loose windmills, each with a solar panel)
5. Fossil fuels	5. Sustainable sources: wind, water, sun

Liveability

1. Vital villages	1. Shrinking villages
2. Center facilities and housing in cores	2. Facilities and living in villages
3. Forced self-reliance	3. Selected self-reliance
4. Major transport arteries (roads, public transport)	4. Fine-mesh structure (roads, public transport)
5. New residents (migrants, international students)	5. Departing residents

Governance

1. Central control	1. Area-oriented/Self-sufficient villages
2. Monopolies	2. Freedom of choice
3. Control	3. Laissez faire/Every man for himself
4. Control/Involvement in organisations	4. Control/Involvement with residents
5. Trust and leave over to others	5. Distrust and control
6. Knowledge and information accessible to a selective group	6. Knowledge and information accessible to everyone

8.2.4.4 Scenario 4: 'We embrace more water and constantly look for small-scale new possibilities'

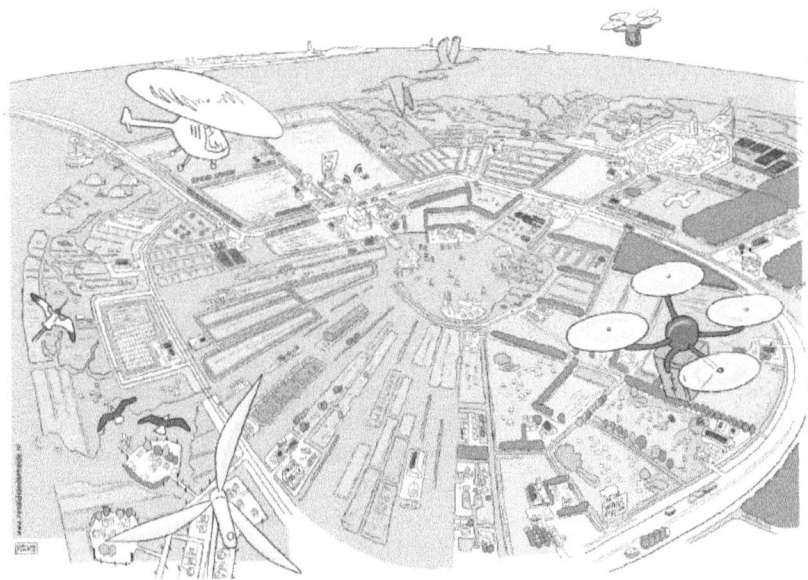

Figure 8.7 Scenario 4
Source: Studio Ronald van der Heide.

Storyline

It is the year 2030. Due to climate change, the amount of water has increased enormously. In addition to high dykes, there are places where the tide has free rein, or where the chances of high water are greater. Friesland is agilely moving along with small-scale, local solutions. Homes are again built on mounds and put on piles. Farmers and local cooperatives regulate water levels themselves and deal dynamically with surpluses or shortages of water. New nature is being created and the varied landscape is a good breeding ground for organic and ecological agriculture. The farmers produce whole new product lines on a small scale, such as new crops, insects or algae. Energy is used economically by investing heavily in limiting waste. Where energy is needed, it is generated locally and stored by local energy chains. Despite rapidly changing weather types, entrepreneurs in tourism and recreation are very well able to quickly meet the needs of visitors. This can be a natural swimming pool in a salt marsh, but also the digital presentation of cultural history via Virtual Reality. The sector is developing rapidly and has, therefore, acquired a strong pop-up character.

Characteristics of scenario 4

The characteristics are derived directly from the results from the round 1 workshops (June 2017) and were used as the basis for the storytelling and animation of this scenario.

- **Landscape:** moves along with the changed environment, dynamic, small-scale, diverse, temporality/pop-up, rising sea level and declining hinterland.
- **Nature:** diversification of the landscape leads to more variety of flora and fauna. Nature is also available in temporary/pop-up form and, therefore, (possibly) in places where you would not normally expect it.
- **Agriculture:** diversification, quick switching (locust farm, algae as biomass, new cultivation), combinations with nature/recreation, closing local chains and developing networks.
- **Water:** terps landscape, moving/parlotting from low-lying areas, farmers regulate the water level themselves, periods of drought/wetting.
- **Recreation:** weather-dependent attractions and activities Since entrepreneurs are used to continuously responding to the changing weather, they are very well able to continue to meet the wishes and requirements of visitors despite climate change. Local environment/authenticity/diversity is cleverly used (natural swimming pool in the salt marsh). Functions in recreation and tourism are strongly focused on rapid development (life cycle is short) and have a strong pop-up character.
- **Coastal defence:** diverse and depending on what is in the hinterland and the extent to which it needs protection. There is no 'just-in-case' approach: there are places where the risks/opportunities/effects are higher than in other places.
- **Energy landscape:** implementation of sustainability at local individual/company level with embracing new techniques. No big players but individual products (zero-on-meter) or small collectives.
- **Technology:** technology is used/applied for new products and forms of production (agriculture), activities (recreation), forms of transport.
- **Governance/management:** continuous search for new developments, opportunities and applications, strongly focused on innovation and less on sticking to the successes of the past. Focus on temporality and pop-up.

Configuration of the 'mixing console' for scenario 4

Nature	
1. Large, contiguous areas	1. Network structure
2. Spatially separating functions	2. Interwoven functions
3. Biodiversity is declining	3. Biodiversity is increasing
4. Dehydration	4. Wetting
5. Healthiness for flora and fauna	5. Human health
6. Value creation: usable raw material	6. Valuable
7. Park landscape ('de-wilding')	7. New wilderness ('re-wilding')

Recreation & tourism

1. Mass	1. Niche
2. Focus, clustering	2. Scattered
3. Large scale ('big attractors')	3. Small scale
4. Permanent	4. Pop-up/Temporary
5. Virtual reality (= place independent)	5. Augmented reality (= place dependent)
6. Routes and arrangements (connecting, marketing)	6. Development of physical supply (Improve, renew, add)
7. Entertainment (= place independent)	7. Area-oriented (= place dependent)
8. Water recreation	8. Recreation on land
9. International market/orientation	9. Local-regional market/orientation

Landscape

1. Hard coastline	1. Soft transition
2. Separating functions	2. Interwoven functions
3. Monotonous landscape	3. Varied landscape
4. Production landscape/Industrial landscape	4. Residential landscape/Experience environment
5. Static landscape	5. Dynamic landscape/Adaptive landscape

Agriculture

1. Diversification/Multifunctional	1. Monoculture/Monofunctional
2. Global market	2. Regional chains
3. New emerging shapes	3. Preserving/Nurturing traditional forms
4. Nature-exclusive	4. Nature included
5. Regeneration: interventions	5. Degeneration: depletion of the soil
6. Extensive agriculture	6. Intensive farming
7. Biological-ecological	7. Industrial (GM)
8. Small scale	8. Large scale
9. Scaling down	9. Scaling up
10. Area cooperatives	10. Chain managers
11. Cottage industries	11. Vacant agricultural buildings

Climate & water

1. Based on technical solutions/processes	1. Starting from natural solutions/processes
2. Central (government)	2. Local (cooperatives, individuals)
3. Big water task	3. Small-scale solutions
4. Dispose of water/Store regionally	4. Retain water/Store locally
5. Preventing consequences	5. Embracing the consequences: rewetting, dehydration, salinisation
6. Fossil fuels	6. Renewable, sustainable resources

Economy	
1. New emerging shapes	1. Preserving/Nurturing traditional forms
2. New professions	2. Disappearing professions
3. Specialised knowledge	3. Diversification
4. Craft	4. Automation/Robotisation
5. Large scale is vulnerable	5. Large scale is efficient
6. Small (SME, family) companies	6. Major international players
7. Scale advantage ends up with citizens	7. The advantage of scale ends up with the company
8. Chain managers	8. Area cooperatives
9. Young people	9. Elderly
10. Small number of large employers	10. Broad base of small SME's

Energy	
1. Slow transition (small steps)	1. Rapid transition (big bang)
2. Dependent on other parties	2. Self-sufficient
3. Revenues to energy companies	3. Revenues to communities (individuals, cooperatives)
4. Large-scale, clustered (wind farms, solar farms)	4. Small-scale, distributed (loose windmills, each with a solar panel)
5. Fossil fuels	5. Sustainable sources: wind, water, sun

Liveability	
1. Vital villages	1. Shrinking villages
2. Center facilities and housing in cores	2. Facilities and living in villages
3. Forced self-reliance	3. Selected self-reliance
4. Major transport arteries (roads, public transport)	4. Fine-mesh structure (roads, public transport)
5. New residents (migrants, international students)	5. Departing residents

Governance	
1. Central control	1. Area-oriented/Self-sufficient villages
2. Monopolies	2. Freedom of choice
3. Control	3. Laissez faire/Every man for himself
4. Control/Involvement in organisations	4. Control/Involvement with residents
5. Trust and leave over to others	5. Distrust and control
6. Knowledge and information accessible to a selective group	6. Knowledge and information accessible to everyone

8.3 Outcome

Het Omgevingslab Fries Platteland has provided a wealth of information. Looking at the common thread in the dialogue, the participants are reasonably in agreement on a number of major themes at a higher level of abstraction. These are topics that emerged in all meetings – the results show that no clear, striking regional differences have been identified within Friesland on these themes.

What is striking is that:

- participants expect that the scale of landscape/companies/governments, technological developments and the way in which we deal with climate change are the most important for the future of the Frisian countryside. Young people do not think very differently about this, but see an increase in tourism, resulting in an increasing tourist flow and increasing demand for space for facilities, as an important topic for the future of the Frisian countryside;
- participants have a concern when it comes to large-scale and upscaling of, among other things, agriculture, multinationals, scaling up (public) organisations;
- this has to do with the feeling of loss of control, vulnerability, distance from the citizen, monotony, separation of functions and landscape pain. There is a desire for a certain degree of small scale for the benefit of landscape diversity that is experienced as quality (in combination with a mix of functions), local authority and influence on the processes that influence the Frisian countryside and its identity;
- participants recognise that large-scale can be efficient and effective, for example, in relation to agriculture for the global market, but not necessarily as the best way or best solution to deal with issues – because large scale also has a degree of dependence and thus vulnerability, there is distance from citizens can arise and it is uncertain whether the countryside can absorb the large scale sufficiently without losing too much (in identity). The question/concern is to what extent scale size and enlargement, and specifically of agriculture, do or do not fit with spatial quality/landscape types/landscape quality/nature/human dimension?
- participants find the dependence on large players (employment) vulnerable and see a stable basis in (family and SME) companies with regional ties;
- participants attach great importance to biodiversity. On the one hand, as a counter-reaction to scaling up and large-scale and to the (perceived) exhaustion and rationalisation of the landscape. On the other hand, for the benefit of meadow birds, identity, landscape quality and attractiveness of the landscape;
- participants experience the identity of places as very important. It is uncertain whether place identity can survive in view of the (macro) socioeconomic, climatological and demographic developments that will affect Friesland. The attractiveness and experience of the landscape is important. Diversity plays a role, but the landscape should not become cluttered (see fear of scaling up and the danger of landscape pain in relation to nature, landscape, agriculture);

- participants often expressed the wish to involve citizens more in interventions that affect the design of their living environment. The expectation is that people will then be more connected to these interventions and that the identity and quality of their own environment will also be strengthened;
- participants show interest in the area cooperative as a vehicle for arranging (regional) activities;
- participants do not have infinite confidence in or sees (financial) support for technical solutions in the fight against the consequences of climate change but prefers them in the short and medium terms, in order to maintain a sense of security and stability for landscape design and use. On the other hand, embracing climate change and, thereby, giving more space to natural processes that can lead to a lot of innovation (new crops, different water management, new target groups for tourism). But there are also concerns about safety;
- participants need clarity about the degree of 'leeway' that exists in policy for entrepreneurs, organisations and social initiatives (frameworks vs. freedom);
- there is a need for knowledge when it comes to the development of and dealing with 'great transitions', such as climate change, developments in agriculture, mobility, demography and shrinkage, and so on. What knowledge is there? Where can this knowledge be found publicly? Which sources do the government parties take as a starting point/truth?
- participants are not unanimous on the most desirable scenario. However, scenario 2 is seen by a majority as the most realistic scenario. The safety offered by reliable coastal defences is experienced as important and small-scale is experienced as appropriate for society, the landscape and the challenges facing Friesland;
- participants have a penchant for small scale and for a dynamic landscape (scenarios 2 and 4);
- little has been said about the topics of the environment, circular economy and technology. It is believed that technology is an important development that influences the future of the Frisian countryside; and
- no clear differences emerge between regions. Perhaps the subjects, dilemmas and type of solutions are not fundamentally different. How one implements solutions in concrete terms can differ. However, that does not result logically from the workshops.

8.4 Concluding Remarks

The Omgevingslab Fries Platteland provided a lot of information for the 'Omgevingsvisies' (environmental visions) for public sector stakeholders. A picture has emerged of the subjects for rural Frisian that should be addressed in environmental visions and which dilemmas should be addressed. A number of subjects – red threads – are particularly striking. The common threads in the discussions in the Lab are two integral dilemmas regarding scale and possibility space.

8.4.1 SCALE: Large scale versus small scale

There is a concern about large scale and scaling up of, among other things, agriculture, multinationals, scaling up (public) organisations. This has to do with the feeling of loss of control, vulnerability, distance from citizens, monotony, industrialisation of the landscape, decrease in biodiversity, separation of functions and landscape pain that leads to loss of identity. There is a desire for a certain degree of small scale (human scale) for the benefit of landscape and biodiversity, which is experienced as quality (in combination with function mixing), local authority and influence over the processes that influence the Frisian countryside and its identity.

8.4.2 POSSIBILITY SPACE: Frameworks/Government versus freedom/society

On the one hand, there is a need for leeway for local entrepreneurs, residents and organisations to develop initiatives, as they see fit. On the other hand, there is a need to draw boundaries through legislation and regulations, for example, to obtain (legal) certainty, to prevent or minimise negative impact, or to stimulate quality. Coordination between frameworks and initiatives – whether or not through area cooperatives – is a precondition in this regard.

8.5 Discussion Questions

(1) When aiming to mobilise large groups of stakeholders to involve a representative sample of the population of a particular tourism destination, how can this be organised optimally? Who to involve and how are roles and responsibilities divided?
(2) To what extent does tourism policy go beyond a focus on branding, marketing, communication and economics?
(3) How to integrate the interests of tourism and recreation into spatial planning processes?
(4) When considering scenarios for the futures of tourism, what combination of for instance a written narrative, pictures (realistic, artistic, abstract), movies, speech, the entourage of the workshop setting works best to get message across to the audience? Do you need to tailor the approach to the specific audience you work with?

Note

(1) World Café is a method for setting up a lively joint dialogue. This method lends itself to both (very) large and fairly small groups. The procedure is as follows. Participants come together in subgroups and discuss the implications and most desirable actions with a view to the future of the sector on the basis of pre-selected themes. The participants work in several short 'work rounds'. The working rounds are short, intensive and focused on creativity. At the end of a work round, participants go to another table to continue embroidering and associating on the things already mentioned and noted on paper sheets/tablecloths. One person chairs the table and stays at the table to explain what the previous group has written. The idea is that people with different backgrounds and expertise meet, talk to each other and look for integrated solutions. The World Café method is a working method in which a group of people takes up the challenge to arrive at jointly supported, concrete proposals.

9 Futures Lab Fryslân

Learning Points
- Strategic foresight and scenario planning could be incorporated in a Futures Lab that is structurally embedded in the organisation.
- A Futures Lab offers the opportunity to think, debate and shape the strategic learning process about the future.
- A Futures Lab offers the opportunity to work across the boundaries of policy domains and disciplines.
- The cyclical character of a Futures Lab allows new parties/stakeholders to join in at each loop.
- Ideally, a Futures Lab is built around a physical and permanently accessible meeting place.
- Software platforms such as Mural, Miro or Microsoft WhiteBoard provide a valuable alternative to a physical environment.[1]

9.1 Background

Friesland is one of 12 provinces in the Netherlands, located in the northern periphery of the country (see Figure 9.1). Friesland is the largest of the 12 with a total area of 5749 km2. Friesland is believed to be part of the remains of the kingdom Magna Frisia that reached from the north of contemporary Belgium to the south of contemporary Denmark. From 1250, Friesland started to develop a distinct political organisation and since the 19th century a separate identity. The Frisian language is officially acknowledged and the main language of the Frisian government and historically, strongly related to English. The landscape of the mainland of the province is featured by agricultural land, lakes and three national parks. The province also covers the western part of the UNESCO Wadden Sea area, including four islands, of which one is a national park. The province has a long coastline along the IJsselmeer and the Wadden Sea. Scattered across the mainland there are many villages and the renowned small to medium sized 'eleven cities'. The city of Leeuwarden is the capital of Friesland and the seat of the provincial government. In the Netherlands, provincial governments represent the administrative layer between the national government and local municipalities. It is for subnational and regional policy issues.

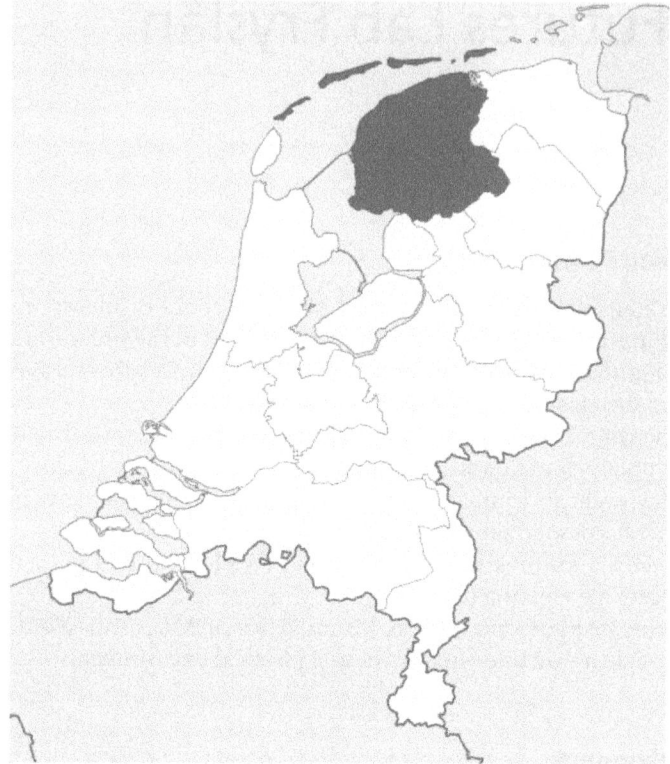

Figure 9.1 Location of the province of Friesland in the Netherlands
Source: Microsoft Office (licensed by CC BY-SA).

> Provinces are governed by the Provincial Council (*Provinciale Staten*), the Provincial Executive (*Gedeputeerde Staten*) and the King's Commissioner. Provincial Councils are comparable to the House of Representatives or the Second Chamber in national governments. They are the highest body of the province and determine what the province does in general terms. The members of the Provincial Council are elected by the inhabitants of the province, The Provincial Executive is the administration of the province and can be comparted to the cabinet of a national government. Its deputies are elected by the Provincial Council. Each deputy deals with one or a few substantive portfolios. The King's Commissioner is the chairman of the Provincial Executive and also the chairman of the Provincial Council. The Commissioner is not elected but appointed by the Crown, which is the King of the Netherlands under ministerial responsibility.[2]
>
> Some of the tasks of provinces are laid down by law. However, according to the constitution, provinces can decide for

themselves which additional tasks they want to tackle. Provinces in the Netherlands are mainly concerned with sustainable spatial development; environment, energy and climate; vital countryside and nature management; regional accessibility; regional economy; culture and monument conservation; and supervision of municipalities within the province.[3] The province of Friesland has 19 municipalities.

The province of Fryslân[4] has two staff departments: strategy and concern control. The organisation is divided into five substantive task areas, each of which is led by a so-called 'task manager'[5]: Green Fryslân, Finance and Resources, Livable Fryslân, Enterprising and Energetic Fryslân, Destination Fryslân. Each of these task areas has a number of staff departments, operational departments and pools, including a number of substantive teams.[6]

For tourism, quietness, space, nature, culture, the Wadden Islands and (water) sports can be considered as the most important attraction values of the province. Four out of five tourists in Friesland (80%) come from the Netherlands, one in five (20%) is a foreign guest. Until the outbreak of COVID-19, the number of foreign and domestic visitors to Friesland grew strongly and it was expected that this growth would continue until 2030. The figures in Table 9.1a show that tourism has fallen sharply due to COVID-19 but seems to be recovering. At the same time, the share of domestic tourism seems to be increasing due to the pandemic. Foreign tourists mainly come from Germany, but the share of German tourists has decreased by 10% due to the COVID-19 pandemic (see Table 9.1b). Compared to the whole of the Netherlands, Friesland

Table 9.1a Number of visitors to province of Friesland, by origin per year (*1000)

Origin	Year		
	2019	2020	2021
Domestic (from Netherlands)	1411 (80%)	1316 (88%)	1588 (92%)
Inbound (from abroad)	361 (20%)	179 (12%)	144 (8%)
Total	1773 (100%)	1495 (100%)	1732 (100%)

Source: CBS Statline, accommodation statistics from 11 July 2022.

Table 9.1b Origin of international tourists to province of Friesland, per year (*1000)

Origin	Year		
	2019	2020	2021
From Germany	266 (74%)	134 (75%)	93 (65%)
From Belgium	39 (11%)	23 (13%)	18 (13%)
From elsewhere	56 (16%)	22 (12%)	33 (23%)
Total	361 (100%)	179 (100%)	144 (100%)

Source: CBS Statline, accommodation statistics from 11 July 2022.

attracts relatively few visitors from other European countries, Asia and America.

Recreation and tourism have a significant economic importance, in terms of spending and employment. The share of employment is 7% and this share is growing steadily. Spending is also increasing in size.[7] The provincial policy, with regard to recreation and tourism in the period 2020–2028, is mainly aimed at finding the right balance in the distribution of tourism, so that it can contribute to the quality of life of the Frisian population.[8] Current recreation and tourism policy responds to three important developments: the consequences of climate change, changes in the composition of the population and the legacy of the European Capital of Culture in 2018. Friesland is the first region in the Netherlands to choose 'broad prosperity' as a starting point. In 2023, this will be adjusted to 'smart growth', in which the happiness of the Frisians must come first. 'Smart growth means focusing on (cultural) tourists who are interested in the quality and identity of our province, more year-round visits instead of peaks and more visits to North, Northeast and Southeast Friesland. It also means innovation to prevent staff shortages. In order to make it interesting for young people to stay in Friesland after their education, extra spending in this sector is expected to generate (young) SME start-ups' (Provincie Friesland, 2019: 41).

9.2 Future Challenges

Today's society is complex, dynamic and quite unpredictable in its development. On the one hand these characteristics are related to the transition that society is currently going through, and on the other hand to the numerous forces that impact upon society, for example in the technological, demographic, economic, social and political domains. This relates not only in the province of Friesland but also in the Netherlands, in Europe and beyond. If an organisation, like the province of Fryslân, wants to sustain a relevant position in the future, the complex and dynamic force field compels the province to look at that world in a different way and to relate to that world in a different way. This requires the province to be resilient and adaptive.

In 2019 the corporate strategy department of the province of Fryslân has started a process to strengthen the strategic capacity of the provincial organisation. One way to achieve this is to start a conversation about the future, in the form of a dialogue, first internally, and then externally, by means of a scenario planning process embedded in the provincial organisation.

9.3 Purpose

The aim of the province of Fryslân (in the remainder of chapter this organisation is referred to as 'the Province') is to build up a story/vision for Fryslân 2030, in a cyclical manner that offers inspiration for starting points for long-term strategies with which todays emerging challenges can be tackled. The story should be easy to tell, stimulate others to think and have the commitment from the provincial government. The story should be based on developments in the Province, in the Netherlands, in Europe and beyond. Although it may include wishful thinking and fantasy, it must be possible to substantiate it with identifiable trends and developments, but without the intention to write a scientific report. The ambition is to realise the story through a joint dialogue that is built up cyclically. The dialogue should start with a small group of interested people from different domains within the Provincial organisation. To enrich the story, the dialogue will be broadened to other parts of the organisation, other governments and organisations in the Province and with the government of the Province. Eventually, a committed structural network will have evolved of engaged participants that shares the story and keeps it up to date. The story provides an understanding of similarities and differences in viewpoints and creates input for shared long-term goals that challenges them, inspires them and stimulates their creativity.

9.3.1 Strategic foresight

In order to develop the story as intended, preference was given to the strategic foresight and scenario planning method, as used by the European Tourism Futures Institute (ETFI) at NHL Stenden University of Applied Sciences, and so the ETFI was asked to facilitate this process.

Strategic Foresight is a competency. It is defined as 'the ability to take a forward view that enables action to be taken with reference to, and within the context of, the future' (Conway, 2007). The European Foresight Platform of the European Union defines it as 'a systematic, participatory, future-intelligence-gathering and medium-to-long-term vision-building process aimed at enabling present-day decisions and mobilising joint actions'. The platform sees Strategic Foresight an open, participatory and action-oriented process in which the participants jointly map, discuss and shape their future[9], as illustrated in Figure 9.2.

This definition of strategic foresight is in line with the so-called *shared and adaptive governance* approach. 'Shared' means that the stakeholders remain continuously involved in the policy process. 'Adaptive' means developing flexibility and resilience to continuously adapt to the changing future (Hartman, 2016, 2018b, 2023).

ETFIs approach to strategic foresight is based on the following principles:
- Thinking about society in complexity and systems. This implies that there is a network of relationships and interdependencies between layers within the organisation (and within government), between

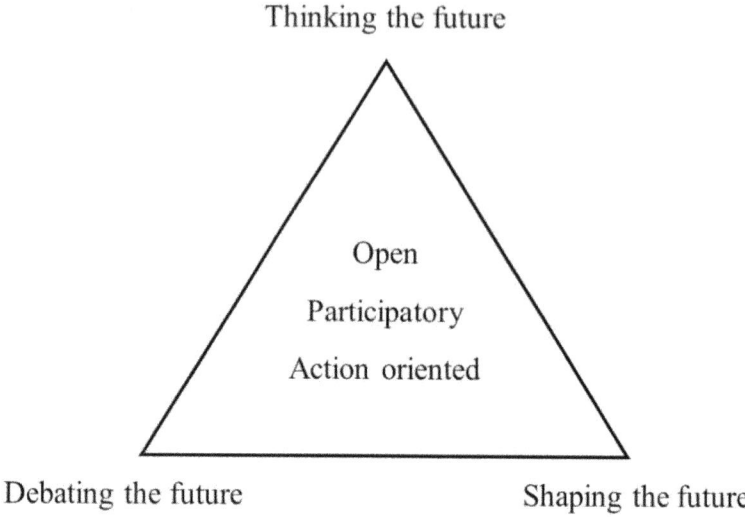

Figure 9.2 Strategic foresight defined
Source: European Foresight Platform (n.d.).

sectors (and policy domains) and between actors (including political regions) that can influence the organisation.

- Thinking about the future: not linear and unambiguous (in statistical forecasts), but non-linear and multiple (in narrative scenarios), taking into account unexpected events (wildcards, black swans, disruptions), shocks (sudden changes), stresses (gradual changes), tipping points and exponential growth or decay of the system. Due to the complexity of the system, policy can also have unexpected consequences.
- Thinking in adaptivity: having the conviction that the organisation is a (sub)system that can develop, evolve and become resilient and future-proof and can adapt to changing circumstances.
- Thinking about the strategic orientation of an organisation: outside instead of inside out. The organisation is connected to a dynamic external environment that constantly creates new opportunities and threats and, therefore, must be constantly monitored, analysed and interpreted.
- Thinking about knowledge development: fed by stakeholders and not by experts, focus on the process and not on the end product. Strategic foresight is a strategic learning process of co-creation that focuses on what is needed to anticipate, reflect on and learn from the feedback of the constantly changing environment, rather than an approach that focuses on the creation of a scientifically sound product (final report) by an expert. In this process, stakeholders[10] bring together their interests, views and experiences, share their knowledge and insight, learn from each other, learn to better understand each other's perspectives, build a relationship of trust

and their adaptive skills to deal with external uncertainties. Through the working methods and the interaction, an attempt is made to stimulate creativity, to break through the dominant thinking, and thus to promote thinking outside the box. Such a way of developing knowledge does require the involvement of the participants or the use of time, money and resources.

9.3.2 The policy lab approach

The ambition was to establish the dialogue, expand the group of participants stepwise to outside the organisation, and to realise the story in the setting of a policy lab. According to the UK government, a policy lab is a creative space that allows policy teams to develop knowledge and policy development skills in a more open, data driven and user-centred way. It makes use of people-centred design approaches to policymaking. The aim is to co-design new and potentially transformative solutions to the intractable, complex, systemic policy challenges of today. The creative space may be physical or online but should be neutral. It should allow for collaboration across departments, with external experts and with the public. It makes use of an agile, flexible and iterative approach. Experiences by the UK Government show that such a policy lab approach can be used for projects across a range of policy areas, as long as it is customised and fundamentally about understanding people better and designing with them and for them. Projects may run for several months up to a year. The UK Government uses the fundamental steps of Appreciative Enquiry (define, discover, develop, deliver) to provide a broad guideline for its projects. Within this framework, methods may be used, such as collaborative workshops of one to three days or speculative design techniques, to think about the future of, for example, rail travel in 2035. The idea of a policy lab is grounded on the idea of open policymaking: being open to new ideas, new ways of working, new insights, new evidence and experts. An open policymaker is curious, networked and collaborative and digitally engaged. The open policymaker talks to a broad range of people and experts, uses the latest analytical techniques and takes an iterative approach to implementation.[11]

In line with the basic ideas of such a policy lab, a cyclic and repetitive process was developed in close cooperation with the corporate strategy department of the Province that allows more parties to join stepwise. It includes the following steps that would be repeated every year:

(1) Organisation and facilitation of human and financial resources by the Province.
(2) Kick-off sessions (once a year), named 'proeverijen' (tasters) (to motivate new participants to get involved).

(3) Horizon scanning, named future scanning here (continually by participants).
(4) Defining the driving forces of future change (four sessions a year).
(5) Development (and in next cycles revisiting and calibrating) future scenarios (twice a year).
(6) Implications for long term strategic policy (twice a year).

The infrastructure to facilitate this process was named Futures Lab Fryslân. During the course of the process details of the subsequent steps were aligned with the needs and possibilities of the Province and adjusted to the limitations posed by the COVID-19 pandemic. The steps and their outcomes are discussed below.

9.3.2.1 Step 1: Organisation and facilitation

The Province started with establishing a group of strategically oriented persons within the organisation. Feeling for the subject and willingness to give a positive contribution to the Futures Lab were key requirements for participation. While the initial group counted 15 participants, eventually, 31 persons working at the Province have joined the group, consisting of strategic advisors, senior policy advisors, programme, project and task managers, team leaders and regional managers. They were considered remarkable people during this new process. Remarkable people are persons who stand out from those around them by the resourcefulness of their minds (Gurdjieff, 1960). They are 'intensively curious but sharp observers, who understand the way the world works and have their finger on the pulse of change' (Van der Heijden *et al.*, 2002).

The group of participants represented a range of policy domains: energy transition, nature development, agriculture and food production, housing, information and facilities, organisation development, economic affairs, public affairs, infrastructure, mobility, digital innovation, cultural and social affairs and the Wadden area.

A concern strategist at the Province took responsibility for the internal planning and coordination of the project, while a senior advisor was asked to be involved as a more creative member of the project team and writer of scenario narratives. Each step required in depth consideration, discussion and fine tuning to make it fit with and relevant for the provincial organisation.

The initial plan was to establish a central place in the organisation with a huge bulletin board, where the participants could place relevant information visible to others in the organisation, with the intention to raise discussion, identify patterns in the observations and challenge others to contribute too. However, due to the COVID-19 outbreak, this was put aside. Instead of dedicating a place within the venue of the Province to meet, share observations and have workshops, COVID-19 forced the Futures Lab to work online. MURAL has been used as an

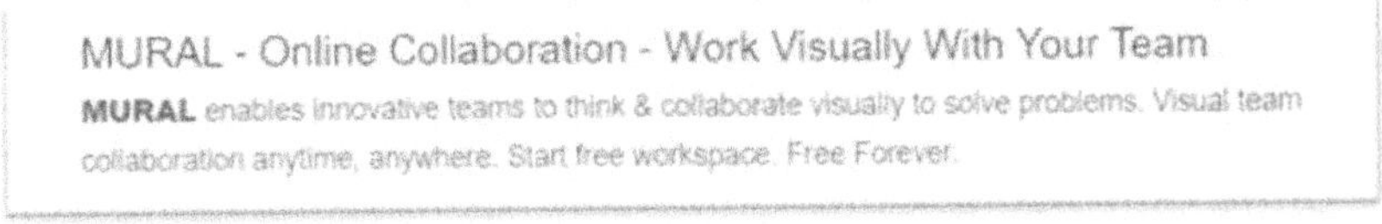

> MURAL - Online Collaboration - Work Visually With Your Team
>
> **MURAL** enables innovative teams to think & collaborate visually to solve problems. Visual team collaboration anytime, anywhere. Start free workspace. Free Forever.

Figure 9.3 Advertisement of MURAL on internet

online collaboration platform to post information and to facilitate online sessions[1], the ETFI designed specific MURAL templates to work with, and MS Teams was used to talk with each other while working in MURAL (see Figure 9.3). In addition, a WhatsApp group was established for day-to-day communication and for sharing relevant newsworthy information among the participants.

9.3.2.2 Step 2: Taster and kick-off sessions

The Futures Lab activities started in November 2019, with two live 'taster' sessions to inform potential participants and motivate them to join. The background and principles of strategic foresight, the process of scenario planning and the aims and ambitions of the project were presented and discussed. The attendees who were willing to join were invited to a kick-off meeting in January 2020, to get acquainted with the first step of horizon scanning (which was named 'Future Scanning' in this project) and the specific task they would have to perform. An exercise was done to get familiarised with how horizon scanning works in practice (see Figures 9.4a and 9.4b).

9.3.2.3 Step 3: Future scanning

Future scanning (the term for horizon scanning that was used in this Futures Lab project) is an analogue to horizon scanning, about systematically monitoring and identifying weak signals as early indications of change (such as, emerging issues, trends or other developments), wild cards, persistent problems, opportunities, risks and threats. Therefore, it refers to gathering relevant information and knowledge about the macro- and meso-environments from a diversity of sources. This may concern items or background articles in the print media, social media or in trade journals; research or policy reports from governments, industry associations, companies or institutions; reports from research firms; blog articles; personal observations of particularities (think of early warnings), etc. The macro-environment is also referred to as the contextual environment of an organisation, which is uncontrollable. It is often referred to with acronyms, such as DESTEP (demographic, economic, social, technological, economic and political changes and developments). The meso-environment acts as the medium through which the organisation is connected with the

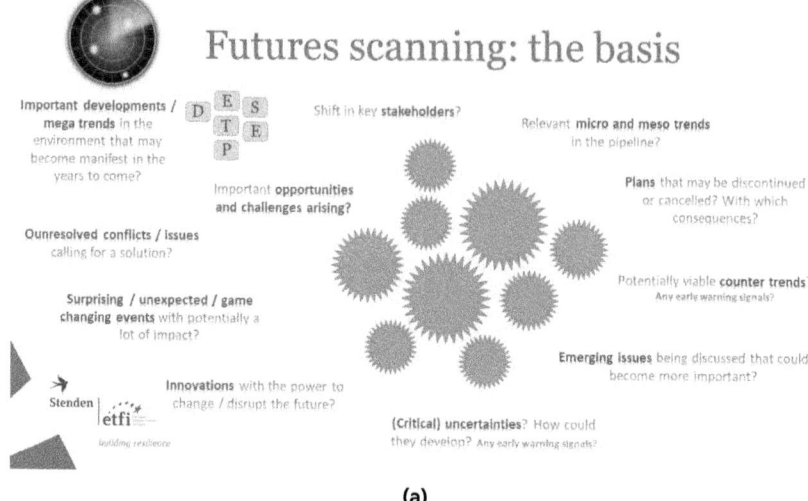

(a)

Features futures scanning process

- Do not seek 'certainty' – it is all about uncertainty
- Be open to what might become, not what is, important for the Province
- Be curious: look beyond the present and what dominates the news today, but look for early indications of future issues and developments that may emerge
- Search beyond the conventional sources, systematically and structured
- Think outside-the-box
- Be aware of cognitive filters and do not let them get in the way
- Test your assumptions continually: why do you find the observation valuable? Do not dismiss observations as nonsense too quickly.

(b)

Figures 9.4a and 9.4b Future scanning instructions

macro-environment. Therefore, it is also referred to as the transactional environment (see Figure 9.5, see also Postma & Papp, 2021).

In line with the aims of the project, a time horizon was chosen until 2030. Based on a few meetings with participants after the kick off, it was decided to take the administrative agreement (*bestuursakkoord*) as a starting point and to designate groups to follow the developments concerning a specific topic: general, energy, green, agricultural, entrepreneurial, liveability, destination Friesland, cultural (including Arcadia 2028, the follow-up of Leeuwarden-Friesland—European Capital of Culture 2018) and togetherness. This prevented that everyone

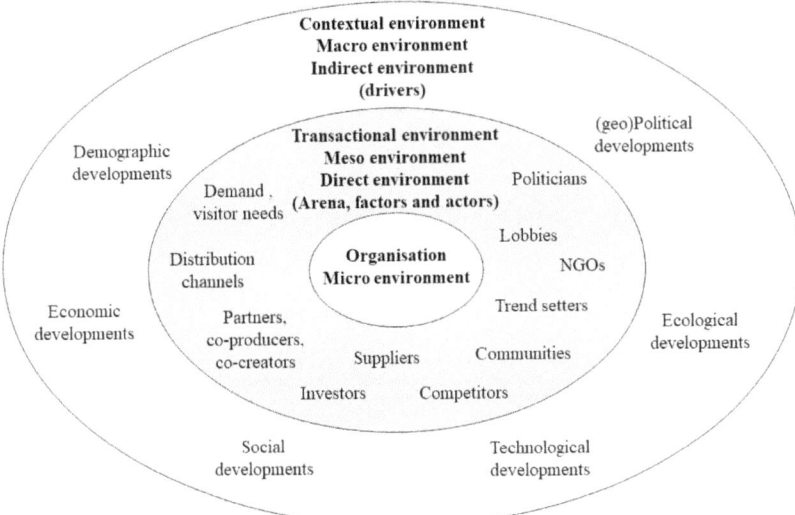

Figure 9.5 The environment of an organisation

would have to monitor any information. To assure motivation and participation, everyone was asked for their commitment and the list of 31 participants was finalised.

The idea of future scanning is that it, along with the cyclic and repetitive nature of the strategic foresight process, eventually develops into a structural activity that employees at the Province do as part of their work, with the intention that they gradually develop a kind of sensitivity for relevant messages, reports, etc. and continually update their understanding of the patterns in the force field at work.

Although the plan was to stepwise involve new participants, also from outside the provincial organisation, there was a concern that the analysis of the results of the future scanning and the development into patterns of cause-and-effect relations would be dependent on the composition of the initial group of participants and difficult to transfer to people who have not been involved in the process in previous rounds.

9.3.2.4 Step 4: Defining the driving forces of change

During a period of eight months (including summer holiday break), the provincial employees involved had been working on their dedicated future scanning. In September 2020, the participants were asked by email to convert and thicken the information they had collected into keywords in a matrix, structured along the demographic, economic, social, technological, ecological and political DESTEP dimensions. The keywords should point at developments that could be advantageous

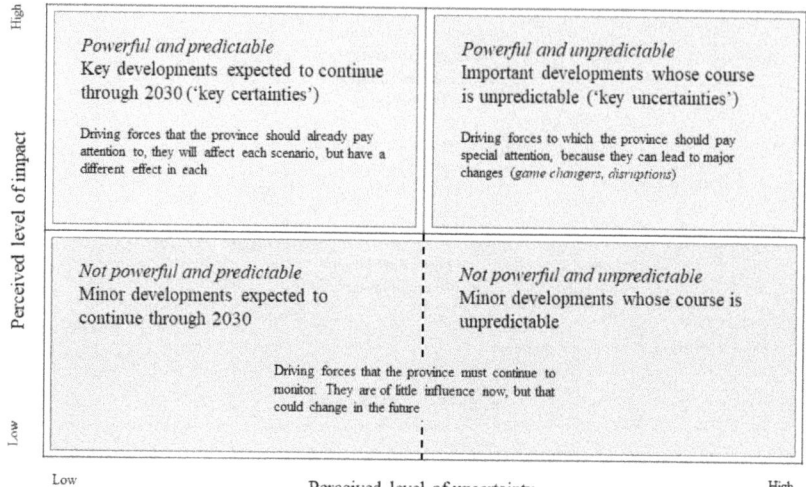

Figure 9.6 Importance by uncertainty matrix

or disadvantageous for the Province. An online session was organised later that month to further thicken and integrate the information into a longlist of drivers that would drive the change until 2030.

During an online session seven weeks later, the drivers of change were reviewed according to their perceived level of impact and heir perceived level of unpredictability/uncertainty (see Figure 9.6). The drivers that were considered most impactful are listed in Table 9.2.

Driving forces considered impactful and certain/predictable at the same time are referred to as key certainties. Driving forces that are perceived to have a relatively strong impact, but whose future

Table 9.2 Long list of driving forces with the highest perceived level of impact

Key certainties (high impact, high certainty)	Key uncertainties (high impact, high uncertainty)
• Individualisation • Focus on sports and health • Changing world order • Change of biodiversity, concern for soil and water • Adaptation to climate change • Transition/transformation of agriculture • Polarisation/dichotomy in society • Increasing power of social media and big internet companies • Rise of populism	• Polarisation/dichotomy in society • Increasing power of social media and big internet companies and information providers • Growing divide between businesses that do and do not adapt to and invest in new technology • Growing divide in society between have's and have not's, active and non-active in economy • Rise of populism and delusion-of-the-day policy • Destruction of democracy (Cycle of Aristotle) • Devaluation of 'truth' • Robotisation and unknown technological developments • Localisation of purchasing behaviour • Transformation of agriculture

Table 9.3 First idea of key uncertainties

Plausible extreme		Driving forces of change		Plausible extreme
The wealthy donate money to please the others	←	*Separation axis:* Polarisation/dichotomy in society	→	The wealthy keep everything for themselves
Algorithm terror	←	*Algorithm axis:* increasing power of social media and big internet companies and information providers	→	Algorithm tamed
Full rejection	←	*Tech axis:* growing divide between businesses; robotisation and unknown technological developments that do and do not adapt to and invest in new technology	→	Full embrace
Full stakeholder	←	*Holder axis:* Growing divide in society between have's and have not's, active and non-active in economy	→	Full shareholder
Strongly populist	←	*Populism axis:* rise of populism and delusion-of-the-day policy	→	Non-populist
Fryslân extremely nationalist	←	*Nation/federation axis:* tenability of current form of democratic governance (Cycle of Aristotle); destruction of democracy	→	Fryslân part of federation (Northern) Europe
Entirely subjective	←	*Truth axis:* devaluation of 'truth'	→	Entirely objective
Purchases entirely local	←	*Purchasing axis:* localisation of purchasing behaviour	→	Purchases entirely online
Farmer in the lead	←	*Agriculture axis:* transformation of agriculture	→	Chain in the lead

development is rather uncertain or unpredictable, are referred to as key uncertainties. The margins of uncertainty are delimited by identifying the possible extremes to which each of the key uncertainties could have developed in 2030 (see Table 9.3).

9.3.2.5 Step 5: Developing the scenarios

In scenario planning, it is common to choose two key uncertainties as dimensions on which the scenarios will be built – preferably the ones with the highest perceived level of impact and the highest level of uncertainty. There are two other considerations for the choice of two key uncertainties from the list – they must be independent, that is, they do not belong to the same DESTEP domain and there is not another, perhaps unnamed driver, driving both. The key uncertainties should not be two sides of the same coin, that is, they should not be two 'scores' on another, perhaps unnamed, driver. The two key uncertainties can be referred to as critical uncertainties if they could cause a transformation of the system when it crosses a threshold or tipping point.

The key certainties are developments that are likely to happen in any future scenario, yet in each of the different future contexts these developments will work out in a different way.

The scenario cross

So, to develop four explorative scenarios, a scenario cross needed to be established that was based on two key uncertainties. While these two key uncertainties constitute the dimensions on which the scenarios would be built, the other key uncertainties and key certainties in Table 9.1 would be incorporated in the scenario narratives.

During a later session, the preliminary outcomes of the analysis were reconsidered and adjusted, and eventually, two uncertainties were deduced that were considered critical: 'dealing with scarcity' and 'dealing with change'. These two critical uncertainties were anticipated to have a major impact on the future of Fryslân and its immediate surroundings.

The way in which the critical uncertainties will evolve through 2030 is quite uncertain. The bandwidth of this uncertainty/unpredictability of the key uncertainties is framed by two extreme, yet plausible, outcomes by the end of the given timeframe (2030). Thinking in extremes stimulates outside-the-box thinking. For 2030, the plausible bandwidth of 'dealing with scarcity' was defined by the extremes of 'linear thinking: driven by Gross National Product (GNP)' and 'circular thinking: driven by sustainability' (Sustainable Development Goals (SDGs)). The bandwidth of 'dealing with change' was defined by 'society embraces change' and 'society denies change'. Plausible means that the extremes could be the logical outcome of a chain of events. The critical uncertainties with their extremes resulted in the following scenario cross.

Thus, the following two scenario axes were formulated:

(1) A scarcity axis that focuses on the way the 'Mienskip' (Frisian word for community) deals with scarcity, ranging from circular to linear.
(2) An axis of change stretching from people embracing change to people rejecting change. In both cases this can be done both actively and passively.

The four scenarios

Each of the four combinations of extremes frames another imaginative future. During an online session, the participants were invited to imagine a macro-context framed by each of the combinations of extremes in the scenario cross and to brainstorm and add contents to each of the quadrants. The brainstorm was facilitated with the following question:

> *Try to imagine an environment that is framed by the two extremes. Try to imagine such a future. What kind of feelings does it evoke? What do you*

sense? Think extreme, yet plausible. Be creative. Use your imagination and phantasy. Follow your own thoughts without looking at what the others write up. There is no right or wrong. Write down your feelings, emotions, ideas, observations and draw images, or add images from the internet.

Ideas could also be based on the key certainties as listed in the first column of Table 9.1. These are likely to occur anyhow yet will work out differently in each scenario.

After the session, an employee at the Province volunteered to integrate the outcomes into a consistent narrative of each scenario with a proper title, that pull the reader into a different future context. Based on feedback by the participants, the titles and texts were adjusted a few times during the months thereafter. To spark the use of the scenarios, they are slightly extreme and they magnify certain issues. Nevertheless, they describe a future that is plausible.

Eventually, each scenario was portrayed as a plausible news item about a random day in 2030. Specific styling features were used to underline the message of each scenario and highlight the differences between them. Each title referred to a specific newspaper in which the article is published. The newspaper article was written in a dedicated font and colour, and the scenarios in the upper half of the scenario cross contained English words in the Dutch narrative to underline the international touch. Images from the internet were used to complete the painted picture (see Figures 9.7 and 9.8).

The scenarios are explorations of alternative (extreme yet plausible) futures. They:

- sufficiently address the two extremes of the key uncertainties on which they are built;
- are internally consistent and do not show overlap. They are able to develop independently from each other;
- are new, original and surprising, and sufficiently different than the present;
- provide new insights to inform choices, thus, offer a perspective for action; and
- are plausible and imaginable. Plausible means that they can logically be deduced from the developments in the present. Imaginable because they do not contain alien elements. Thus, the Province would be able to immerse itself into the scenarios because they:
    - describe the environment external to the Province which is thus uncontrollable; and
    - match the relevant urgencies, needs, fears and wishes of the Province and communicate a sense of urgency.

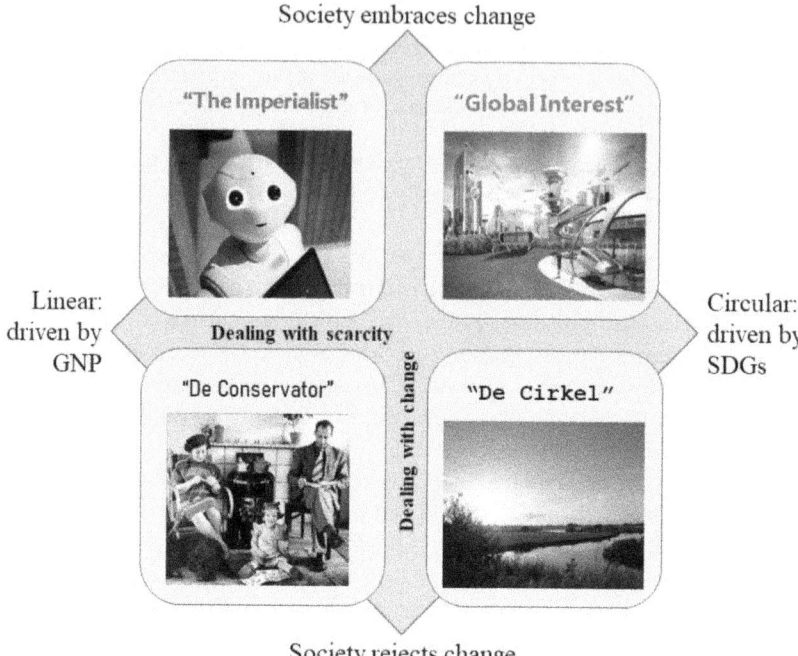

Figure 9.7 Scenario framework with illustrations[12]

9.3.2.6 Step 6: Policy implications of the scenarios

The scenarios are not predictions, however, each one paints an extreme and plausible alternative picture of the world in 2030. Their intention is to raise a discussion about the future and to find inspiration for new policies or evaluation criteria of existing policies.

Familiarisation with the scenarios

To spark the discussion about the future, a session was organised to familiarise the participants with the scenarios. A few stakeholders outside the Province were also invited to participate. To prepare for the session, each scenario was summarised under the same denominators as shown in Figure 9.9. For each of the scenarios three key features were formulated (see Figure 9.10). The features are rather extreme, diverse, without overlap and together provide a broad spectrum of key characteristics of the future context that Fryslân may face in 2030.

The session started with a general discussion of the scenario cross, with the help of the following questions:

(1) Which scenario do you conder the most likely? Why?
(2) Which scenario do you consider the most preferable? Why?

"The Imperialist"

Daily breaking global news and infotainment

1 november 2030

(printed in blue, font Dubai Medium)

"Global Interest"

digitale dagdromen en informatie voor allen

1 november 2030

(printed in green, font Ebrima)

"De Cirkel"

geregeld krantje voor ongeregeld zootje (regio Noordkust)

1 november 2030

(printed in black, font Courier New)

"De Conservator"

Nationale Dagelijkse Courant

1 november 2030

(printed in black, Bahnschrift Semilight Semicondensed)

Figure 9.8 A taste of the four scenarios: Headlines of four newspaper articles

(3) Is there a difference between the most likely and most preferable scenario? What does that mean?

As soon as all participants were a bit more immersed into the alternative futures and were familiar with the key differences, the following questions were used to discuss the scenarios one by one:

(1) Which emotions does the scenario generate?
(2) What are the strengths and weaknesses of the scenario?
(3) What do you hope for, what do you fear in this scenario?
(4) How does the Province look like in the scenario? What role does it play?

MURAL templates were designed to facilitate the discussion and allow the participants to put their thoughts on virtual stickies.

The participants expressed that they really enjoyed participating in the discussion of the scenarios. The Provincial Government

The Imperialist progressive, business-like and internationally oriented	Global Interest A society driven by ideals, imagination is important
Politics: imperialism and meritocracy, leaders based on merits	Politics: utilism, an ethical movement that measures the moral value of its contribution to the common good, which includes well being and happiness of all people
Power: central, charismatic leader, business	Power: invisible, network, feminine
Cooperation: internationally orientated, maximum cooperation aimed at economic growth	Cooperation: globally oriented, maximum cooperation aimed at ecological growth
Longing for: Q1 (when business results are published) and future	Longing for: 2050 and world peace
Other features: pro active, goal-driven, elitist, feasibility thinking	Other features: pro-active, goal-driven elitist, feasibility thinking, haute culture, circular, technologically green solution, liberal ethics, central position of women
Dominant political parties: VVD, CDA, Google and Facebook	Dominant political parties: D66, Groen Links
Insights colour: red	Insights colour: yellow
World view: liberal, market forces, merchant attitude, VOC mentality	World view: spirituality, agnosticism, esoteric, Buddhism
De Cirkel a local society where local customs and mutual relationships are important.	**De Conservator** A conservative course, in which traditional Dutch values are central. It is also important to nurture the traditional economy. Outside interference is not appreciated.
Politics: federal, tribalism (power of the tribe)	Politics: nationalism
Power: strong federal leadership and very strong local power	Power: central, strong government and strong leader
Cooperation: locally oriented with other circles because of the benefits for the own one	Cooperation: nationally oriented, minimal cooperation aimed at benefits for the Netherlands
Longing for: 1970s	Longing for: nostalgia, 1950s
Other features: reactive, free-spirited, sustainable with former technology and solutions, sober, safe own circle, protection own circle, local production	Other features: reactive, conservation is important, identity of Holland, orderly, simple, protectionist, economy driven
Dominant political parties: SP, Plaatselijk Belang	Dominant political parties: PVV, FvD
Insights colour: green	Insights colour: blue
World view: atheism, belief, mother Earth	World view: traditional western Christian religion, atheism

Figure 9.9 Summary of each scenario in its own font type

also valued the work that was done. Nevertheless, the sessions also showed that it was not always easy to step out of the frame of the current political agenda and imagine and talk about something entirely different.

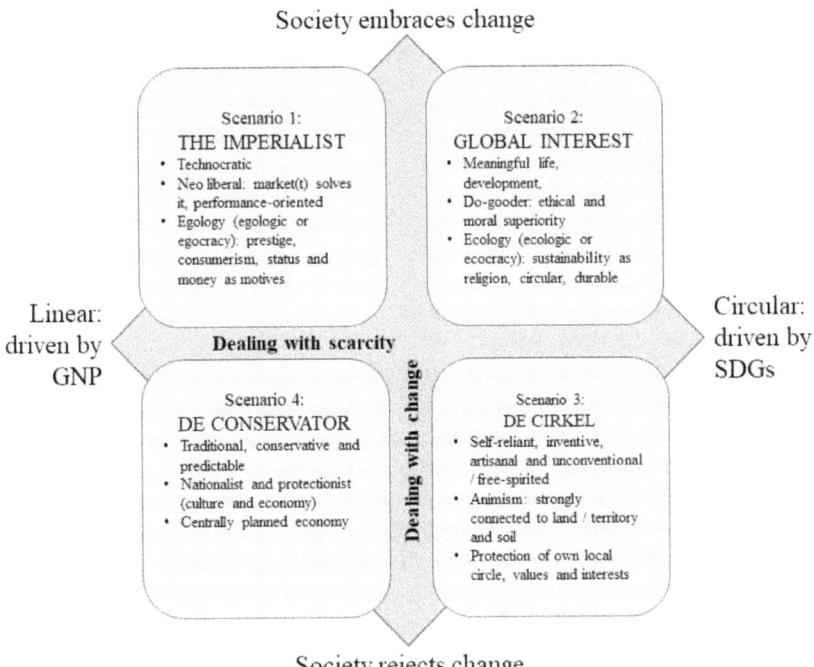

Figure 9.10 Key features of each scenario

Implications of the scenarios

Once the scenarios had been discussed intensively and the participants were familiarised with them, they were able to explore how key certainties (ongoing developments, such as climate change and ageing, as listed in column 1 of Table 9.2) and other key uncertainties work out in each scenario, and the possible inferences of each scenario with relevant policy domains. Eventually this could be used to formulate strategies with which these situations can be anticipated and how some implications can be seized and how others can be prevented. Such strategies are more pro-active and less swayed by the issues of the day.

Each scenario represents a different context that impacts on the Province. For each of the policy domains of Frisian – identity, agriculture, landscape and nature, mobility and housing – a dedicated (online) session was provided to explore the implications of each scenario. The participants received a special invitation, in line with the styling of the scenarios, and a fact sheet with the key features of each scenario. For each policy domain session, a small group of involved participants was invited.

The key features of the scenarios contexts were used to do a structured brainstorm with the participants about possible implications for each of the policy domains. This was done by means of a so-called futures wheel or implication tree on templates within MURAL that were designed by ETFI. Three key features of each scenario were put in the middle. First, ideas for direct implications were generated for each scenario. Next, these were used to create second order implications. This approach, in which the (external) context of the scenario is projected onto the (internal) organisation, is referred to as outside-in thinking.

Consolidation of opportunities and threats

Once the futures wheels were completed, the outcomes were reorganised per scenario. The implications of all policy domains were listed together per scenario on a MURAL template. During four subsequent weeks, an online session was provided for each scenario. Per session, each of the participants had to mark three important implications that they would like to have realised by 2030, and three implications that they would like to have prevented by 2030 for each policy domain. In fact, these 'likes' and 'dislikes' represent opportunities and threats of the envisaged futures. Based on the total number of 'votes', a top five could be established of likes (opportunities) and dislikes (threats) for each scenario and each policy domain.

Afterwards, the corporate strategy department of the province of Fryslân interpreted the outcomes. For each policy domain, a factsheet was written that included the major challenges of each scenario, plus an overarching and so robust strategy to address the challenges of the scenarios altogether. During online sessions in September and October 2021, the factsheets were discussed with the participants.

9.4 Concluding Remarks

At the time of writing, no new actions have been taken. It was proposed that the main outcomes would be shared and discussed with the task managers and with networkers at the Province. The participants were also asked to think about the messages they would like to share with a new Provincial Board of Governors that would be elected in 2023.

The original plan was to repeat the aforementioned steps 1 to 6 twice a year. Knowledge and information about the macro- and meso-environment would be continuously maintained and registered. Every six months, the evolving patterns/clusters and drivers would be calibrated, twice a year the scenarios and the implications for policy would be examined. This to keep the focus on the future. The physical 'wall' would serve as a living source of information and understanding of what is happening in the environment.

Thus, in a small committee, participants of the provincial organisation would gradually work on the competence to master developments in society together, to form a picture of the complex interrelationships in the force field, to translate this together into a number of future scenarios and existing policy together or to develop new policy. Due to the learning process (strategic learning, organisational learning) that the participants go through, they are able to act as ambassadors and driving forces during every second session, also involving others from outside the provincial organisation. The first year focus was on development, year 3 on institutionalisation with the following years focusing on consolidation. Throughout this process, the ETFI would gradually withdraw. In year 4 and beyond, the cycle would have to continue according to the same steps as in years 2 and 3, while the participants would be able to carry out the process themselves. The plan was that, throughout the years, the circle of participants would be extended outside the Province and would eventually also involve citizens in thinking along.

However, due to the COVID-19 pandemic, the process has taken longer than expected and to repeat the process twice a year has not been realistic. So far, it is unclear how and with which speed the process will proceed.

9.5 Discussion Questions

If you consider your own business, organisation or destination:

(1) Which black swan(s) could seriously harm your business, organisation or destination? How would you be alerted in time?
(2) Do you inform yourself actively about trends and developments that could have an impact upon your business, organisation or destination? How?
(3) Do you exchange, share and or discuss your knowledge and expertise about trends and developments with others, other disciplines, and/or other policy domains within and outside your business, organisation or destination? Why? Why not?
(4) How could a Futures Lab help your organisation to become more resilient and future proof?

Notes

(1) Alternative software platforms are available, for example, Miro or Microsoft WhiteBoard.
(2) https://prodemos.nl/kennis-en-debat/publicaties/informatie-over-politiek/de-provincie/de-organisatie/.
(3) https://prodemos.nl/kennis-en-debat/publicaties/informatie-over-politiek/de-provincie/wat-doet-de-provincie/.
(4) In this chapter, Friesland will refer to the province as a region, Fryslân (the translation of Friesland into the Frisian language) refers to the provincial organisation but will be shortened to the Province.

(5) These persons are responsible for the management of the four urgent and integral tasks ('opdrachten') that the Province formulated in its Environmental Vision ('Omgevingsvisie'), in order to certify the basic quality of the living environment (https://www.fryslan.frl/hoe-staat-het-met-de-omgevingsvisie-fryslan; https://youtu.be/LohaB7GmGz0).
(6) Organogram, organogram Provincie Fryslân. https://cuatro.sim-cdn.nl/fryslan/uploads/organogram-frysk_0.pdf.
(7) https://www.fsp.nl/wp-content/uploads/2020/02/FSP2020_publicatie_toerisme_in_FRL-DEF.pdf.
(8) https://cuatro.sim-cdn.nl/fryslan/uploads/Beleidsnota Gastvrij Fryslan 2028 NL.pdf and https://cuatro.sim-cdn.nl/fryslan/uploads/Uitvoeringsprogramma Gastvrij Frysl%C3%A2n.pdf.
(9) http://www.foresight-platform.eu/community/forlearn/what-is-foresight/.
(10) Ideally from the quadruple helix: public bodies, private sector, knowledge institutions and the community.
(11) https://openpolicy.blog.gov.uk/.
(12) Image The Imperialist – free to use stock photo from https://www.pexels.com/photo/high-angle-photo-of-robot-2599244/; Image Global Interest – free download from https://mrwallpaper.com/wallpapers/science-fiction-futuristic-city-5qr81x1by5cy-w9rn.html (haven't been able to get in touch, no response to communication); Image De Cirkel, photograph taken by the author; Image De Conservator: from collection 'Fotografen De Spaarnestad', photograph taken in 1954. Published with permission.

10 Scenarios for Inbound Tourism to the Netherlands – Case Study

Learning Points

- Presently, the vision and strategies to manage inbound tourism to the Netherlands largely rely on quantitative and 'accurate' forecasts.
- The forecast for 2030 was built upon a baseline scenario in which stable and coherent developments in the current era were considered to continue.
- The foresight was complemented by painting alternative futures brought about by unpredictable yet plausible disruptions.

10.1 Purpose

The purpose of the project was to develop a number of future scenarios that would enable NBTC Holland Marketing to identify important uncertainties in the market and the bandwidth within which inbound overnight tourism might develop in the longer term. The scenarios would need to help the organisation to make estimations of the size of inbound tourism up to 2030, both in total and specified per country of origin, according to the country of origin. Unique to this case was the use of two types of scenarios. Firstly, a baseline scenario was developed based on rather predictable drivers of change, Secondly, alternative scenarios were painted which were based on key uncertainties.

10.2 Background

NBTC Holland Marketing is the national Destination Management Organisation of the Netherlands, responsible for the development, branding and marketing of the Netherlands as a tourism destination. Every five years, the NBTC publishes a future perspective of inbound tourism to the Netherlands (NBTC Holland Marketing, 2008, 2013a). The document shows the major trends in the market for international tourism to the Netherlands and pictures the growth perspective that the Netherlands would be able to realise under certain conditions. The reports are intended to provide leads for the development of the Netherlands as an international tourist destination in the longer term. It calls for a pro-active attitude towards inbound tourism and provides the basis for policy development to be reported in Holland Branding & Marketing Strategy (NBTC Holland Marketing, 2013b, 2015) and consequential marketing initiatives. Usually, the NBTC involves various stakeholders from within and outside tourism in the development of these future perspectives.

The previous report, *Destinatie Holland 2025* (NBTC Holland Marketing, 2013a), which was published in 2013, needed an update. More than in the past, the new vision should focus on the implications for tourism product supply that is required to meet the expected demand. NBTC Holland Marketing asked the European Tourism Futures Institute (ETFI), as one of its stakeholders, to use its expertise and approach to develop scenarios to serve as input for the development of (better) estimates of the size of inbound tourism to the Netherlands until 2030, and to identify uncertainties and band widths.

The study was executed by the ETFI in close cooperation with the management, marketing and research team of NBTC Holland Marketing. All steps described below were conducted during live sessions and workshops. Unfortunately, the ETFI did not get the opportunity to involve other stakeholders, as they were already being consulted by the NBTC for other activities concerning the new vision on tourism for 2030. Any overload and confusion was to be prevented.

10.3 Approach

The study, as described in this, follows the steps as described in Chapter 6. The process started with formulating the scenario question, together with the client, during the preparation of the project and the proposal phase.

How will inbound overnight tourism to destination Netherlands develop in the period to 2030?

After the project started, the domain was analysed and mapped out to identify the features of the subject that would need to be included in the outcomes of the study. The domain analysis was required to define the scope and boundaries, time horizon and issues to be addressed (see Chapter 6). The domain analysis was complemented with statistics to take a snapshot and freeze the development of inbound tourism in time. Next an era analysis was performed with the aim to identifying successive periods of relative stability and coherence concerning inbound tourism and the disruptive forces that initiated each of the subsequent eras. Such an era analysis helps to get a proper understanding of the current epoch of inbound tourism to the Netherlands, the disruptive powers that initiated it and the driving forces that are assumed to maintain a stable and coherent development of the present era. A study of these perceived 'certainties' resulted in a baseline scenario of inbound tourism until 2030: the expected future of no surprises would occur. To be prepared for discontinuation of the baseline scenario, alternative scenarios were also developed. The horizon was scanned for early signals, and trends and developments that could lead to other direction of development. The outcomes of the horizon scan were analysed and two key uncertainties were identified. By defining two extremes for both key uncertainties, a framework was established for four alternative futures: Global Village, Hostility, Resistance and Resilience. Together with the outcomes of other research activities executed by NBTC Holland Marketing, the scenarios were used to make statistical forecasts for the future development of inbound tourism to the Netherlands until 2030.

10.3.1 Domain analysis and domain map

The scenario study focuses on the inbound overnight stay tourism. This domain was mapped out during several workshop sessions with the research department of the NBTC. In this section, the domain is specified and delineated using a definition of the subject for which the results are intended, the geographical scope, the time horizon, a map of key and secondary aspects to be included and the challenges the domain is faced with. Such a domain analysis and definition serves to focus the study and not to spend time on irrelevant matters. Table 10.1 describes the domain. Below this is the map that indicates which parts belong to the domain and which do not (see Figure 10.1).

Table 10.1 Domain analysis of the study

Element	Description
Definition of the domain (topic of the foresight)	Inbound overnight stay tourism
Client (for whom the project is intended)	Building block for *Deltaplan Toerisme* (later called *Perspectief 2030*, NBTC Holland Marketing, 2019)
Geographical scope	Netherlands
Time horizon	2030
Domain map (delimitation, categories, what's in and what's out)	See Figure 10.1.
Key questions/issues? (which questions/issues may become crucial for the persistence of inbound tourism)	Balance between positive and negative implications. Negative sentiment/attitude towards tourism. Climate agreement. Environmental impact (in pressure and Euros). Capacity of Schiphol Airport, hotel capacity in Amsterdam. Social visitor pressure. Geopolitical stability/Brexit. Terrorism. Development of wealth/consumer confidence. Tourism as a goal or a means.

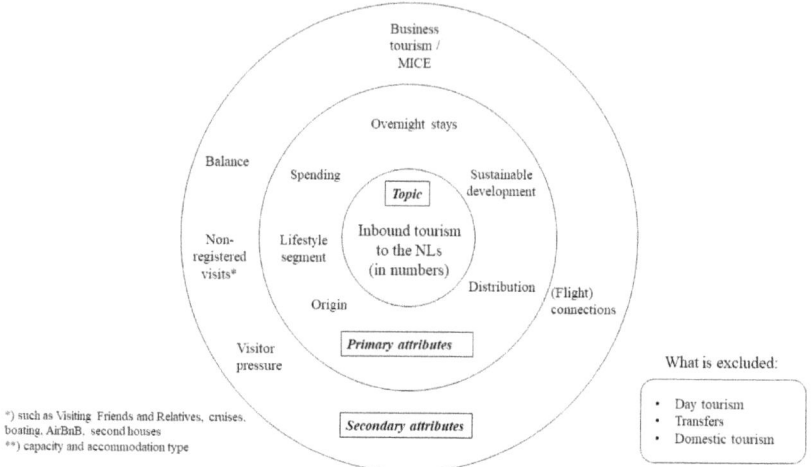

Figure 10.1 Domain map of the study

Domain description and research question serve as a guideline for the implementation of the study.

10.3.1.1 The domain in figures

Based on desk research by NBTC Holland Marketing, the domain could be specified in figures as follows:

- International tourism is experiencing strong growth worldwide.
- Inbound overnight stay tourism to the Netherlands is also growing. This growth is mainly caused by growing economies, a weak Euro, the image of the Netherlands as a safe destination, extra flight connections and the growth of the short holiday market in Belgium and Germany.
- According to the most recent figures from the 2014 Inbound Tourism Survey conducted by NBTC, it appears that inbound residential tourism accounts for more than two thirds (68%) of inbound tourists. The rest (26%) visit the Netherlands for business reasons. Almost three quarters of inbound tourists (73%) travels individually, the rest organised.
- All countries of origin show clear growth from 2012 to 2017, but the fastest growers are the neighbouring countries of Germany (+62%), Belgium (+46%) and United Kingdom (+32%) and Asia (+71%). Growth from other countries and regions is well above 20% in all cases.
- Airplanes (44%) and cars (40%) are the most favoured among incoming tourists. Eight percent come by train, 3% by coach and 4% by other modes of transport. The share of modes of transport has been stable over the years.

- Most foreign stay-over tourists come to the Netherlands for a city trip (35%) or a holiday at the coast (22%).
- This is reflected in how they spread across the Netherlands. Amsterdam has a share of 38%, the rest of the Randstad 21%, the coast 15% and the rest of the Netherlands 27%. Between 2014 and 2017, the distribution of international visitors across the Netherlands remained relatively stable.
- With regard to the type of accommodation, between 2012 and 2017 holiday parks show the fastest rise in popularity (+47%), followed at some distance by hotels (+28%) and camping areas (+25%). Hotels/guesthouses are most popular for a short break of one to three nights (68%) and a holiday park/cottage for a stay of four to seven nights (59%). Campsites appear to be the most attractive for a longer stay (38%).
- More than a quarter of foreign tourists come to the Netherlands for the first time (27%), 21% have been to the Netherlands more than three years ago and 52% less than three years ago.
- Over the years, it appears that transport and accommodation are more often booked in advance and at the same time.
- When inbound tourism is expressed in monetary terms, it appears that the average holidaymaker spends €599 per stay (versus €1035 for business travellers). In total, inbound overnight stay tourism resulted in a spending level of €12.9 billion in 2017, €11.8 billion of which benefits the Dutch economy. Since a job is generated for every €120,000 spent, the inbound residential tourism directly creates 98,000 jobs.

10.3.1.2 Forces that drive the domain

During one of the workshops, a scan was performed of factors that drive the future of inbound tourism to the Netherlands. This resulted in 92 different factors that were clustered into 17 driving forces (Table 10.2).

10.3.1.3 Era analysis

An era is a period of relative stability and coherence. An era usually starts and ends with a disruption. In between is characterised by balanced conditions and changes that only take place gradually. Thus, each era has an identity of its own and can be called 'the era of the ...', depending on what is considered dominant.

In an era analysis, successive epochs are mapped out. The aim is to understand how the current period of inbound tourism can be identified and characterised, which factors have marked the beginning of this era and the circumstances and stakeholders that play a role in it. The outcome of the era analysis determines the context of the remainder of this scenario study.

Based on statistics provided by NBTC, and the scanning of a large number of sources, five eras could be established, as illustrated in Figure 10.2.

Table 10.2 Overview of the forcefield that impacts upon inbound overnight stay tourism to the Netherlands

Travel to the Netherlands	Accessibility of the Netherlands by plane and fast train (such as Eurostar), travel time to the Netherlands, flight capacity of Schiphol Airport and regional airports.
Price level	Price of transport (including flight tickets), accommodation, entertainment, hospitality) and the exchange rate of the Euro.
Branding	History of the Netherlands related to VOC trade and craftsmanship, image and the brand of the Netherlands, branding, marketing and promotion.
International businesses	International businesses.
Hospitable residents	Hospitality and openness of Dutch population and language skills.
Digital infrastructure	Online infrastructure and online information about the Netherlands (such as, results in Google) and the mobile network.
Tourism policy and implementing bodies	Trade missions, integral national tourism policy; NBTC, Tourist Information Centres (VVVs), DMOs.
Carrying capacity/ balance	Residents with an open attitude towards tourism, image of foreigners, distribution of tourists, social and economic carrying capacity, visitor pressure, balance city versus nature.
Payment options	Banks, payment systems and payment opportunities with credit cards.
Free travel across borders in EU	Free movement of people within the EU, open or closed borders (border controls), visa rules, accessibility of the Netherlands (visa, passports), and the Euro.
Overnight stay accommodation	Capacity of the right quality/level of hotels, bungalow parks, camping sites and Airbnb.
Laws and regulations	Governance and degree of organisation Destination Netherlands, policy regulations, facilitation by government, Dutch drugs policy, Ministry of Foreign Affairs/visa, DMCs/PCOs, safety, Alderstafel.*
Booking Facilitators	Offline and online accessibility of provision and booking possibilities (tour operators, booking platforms, online travel agencies (OTAs)), Tripadvisor.
Job market	Labour supply and language skills of Dutch businesses.
Entertainment	Amount, variation and location (also outside Amsterdam) of the supply of tourism products, visitor attractions, stories and experiences.
Business (travel)/offer	Conference locations and conference associations.
Travel within the Netherlands	Accessibility, infrastructure, national and international train connections, transport companies, such as NS and Flixbus, parking capacity and parking fees for cars and tour buses, public transport, electric transport and transport logistics.

*The Alderstafel is a discussion table about the development of aviation in its environment. Established in December 2006 to advise the cabinet on the development of Schiphol Airport in conjunction with Eindhoven and Lelystad airports.

2016–2030 – The (current) era of 'accelerated growth'

Under the influence of the economic growth of the middle class in other parts of the world, such as China, the cheap flights of the LCCs and the growing supply of cheap Airbnb accommodations, inbound tourism is growing faster than expected (an average of 7% since 2010) and is starting to show indications of massification. Amsterdam is becoming full and, due to the scarcity of hotel accommodation, room prices are increasing, which encourages the further growth of Airbnb. The social and spatial capacity of certain places in the Netherlands as

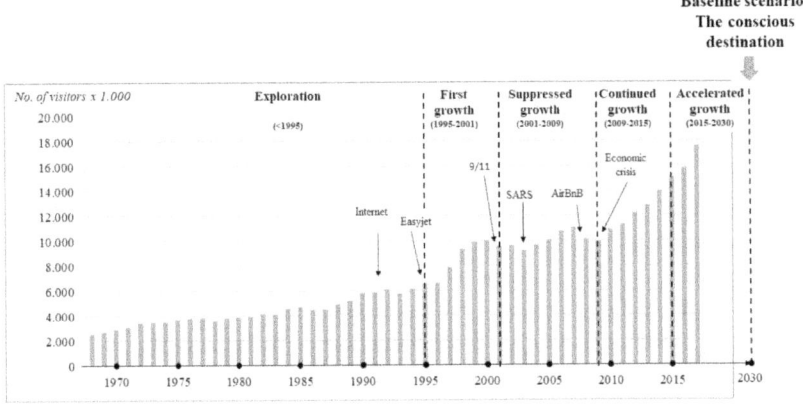

Figure 10.2 Era analysis of inbound tourism to the Netherlands
Source: CBS/NBTC.

a destination is exceeded. The irritation among residents (and other tourists) is growing and a discussion is underway about how the growth of inbound tourism can be better controlled. Plans are being made to spread tourists over other parts of the Netherlands, because on the other hand, many regions and cities in the Netherlands need more visitors.

Main triggers ('disruptors')
- End of global economic crisis, increasing prosperity.
- Increase in global connectivity, especially through low-cost carrier (LCCs).
- Power of online travel agencies (OTAs).

10.3.1.4 Expected developments in the current era

The present era, as described above, began in about 2016. In order to understand how this era is *expected* to develop over the next 12 years, we must, as it were, use the driving forces and growth factors to set the snapshot of the domain in motion. To this end, the NBTC's research department has conducted a study of the 'certainties' for the current era: existing trends, forecasts, projections, constants, cycles and plans in the pipeline. In summary the following picture emerges:

- Growing inbound tourism via air, car, rail and water.
- The growth of inbound tourism (and associated tourist pressure) to Amsterdam will initially increase but will level off after 2024 (in terms of numbers and length of stay) as a result of restrictive measures.
- Growing inbound tourism outside Amsterdam in two 'rings'/zones.

10.3.1.5 Baseline scenario: The conscious destination

The developments of the more or less predictable forces described in the previous section led to a future scenario for 2030 that is called the probable or expected future. This future scenario is presented as the baseline scenario for 2030. It describes how the domain is expected to have developed in 2030.

> It is 2030.
>
> **Number:** Because of the impact of the end of the global economic recession (2015), demographic trends and price developments in aviation, international tourism has grown at an accelerated rate and the number of inbound overnight tourists visiting the Netherlands has also reached a new record. Although the growth can mainly be attributed to air traffic, forms of transport have also clearly contributed to the growth.
>
> **Origin:** Despite the growth in Asian visitors, most tourists still come from Europe, with Germany, Belgium and the United Kingdom leading the way.
>
> **Spread:** Because Amsterdam is physically and socially at its limits, growth has come to a halt and inbound overnight tourism has spread more over the country under the influence of consciously taken measures. The ring road around the Randstad has especially benefited from this, but the more peripheral areas in the country have also seen the number of foreign guests grow in recent years, partly stimulated by an increasingly better infrastructure.
>
> **Overnight stays:** Overnight stays are still predominantly in hotels, but these have faced fierce competition from Airbnb and holiday parks. The length of stay in Amsterdam has levelled off due to a lack of capacity, partly caused by government intervention. The length of stay outside Amsterdam has increased, especially in (more luxurious) hotels, Airbnb and holiday parks.
>
> **Type/Lifestyle:** Two thirds of visitors still come for a vacation, most others for a business visit.
>
> **Expenditure:** No information.
>
> **Sustainable development:** Partly due to the increasing pressure from tourism on Amsterdam in particular, but also due to the increased awareness in society that we are ourselves responsible for the future of our planet, a turning point emerged at the end of the previous decade in the thinking among policymakers, consumers and residents alike. As a result, the Netherlands has developed into a 'conscious destination' that is more balanced than over ten years ago. The interests of residents, companies and visitors, and of different regions, are better balanced in order to ensure sustainable growth of inbound tourism. Also, the visitor has become increasingly aware of the downsides of tourism, and they have increasingly turned away from the hotspots and are looking for authenticity, experience and safety.

The baseline scenario assumes that there will be no special surprises, discontinuities or disruptions in the trends and developments that dominate the evolvement of inbound tourism in the current era. Nevertheless, it is unrealistic to assume that the developments will be stable and coherent until 2030. It is likely that various kinds of causes could lead to discontinuity of the baseline scenario. For example, specific events, or current issues that will be resolved in the coming years, innovations that break through, small counter movements that become mainstream faster than expected, new issues that will start to dominate the political agenda, the emergence of new ideas or paradigms, the unexpected rise or decline of the power and influence of certain stakeholders actors, etc. The discontinuity caused by such factors will likely herald an era of inbound overnight stay tourism that deviates from the baseline scenario. It is, therefore, fairly certain that this baseline scenario will not take place and that an alternative future will have emerged in 2030.

10.3.1.6 Key uncertainties

While the baseline scenario describes the *expected* future of inbound tourism when no surprises occur, the alternative scenarios focus on what could happen when such disruption do occur. The alternative scenarios presented in this chapter have been stepwise developed in collaboration between employees of the ETFI, the management and research team of the NBTC, and employees of Fronteer, a consultancy bureau specialised in strategy development. The external force field that could disrupt the baseline scenario has been jointly explored by means of horizon scanning. A joint analysis and interpretation of the findings has led to the identification of two key uncertainties. These are the forces that will have a major impact on inbound overnight stay tourism until 2030 and at the same time create a lot of uncertainty because the nature and degree of its development and the consequences, thereof, are unpredictable. The unpredictability is expressed by identifying two plausible extremes for both uncertainties. The team identified the following two key uncertainties and plausible extremes:

Disruption: Political-economic disruption versus ecological disruption

There are various threats worldwide that can disrupt international tourism and, therefore, inbound tourism to the Netherlands.

On the one hand, political–economic threats are considered an extreme. Firstly, geopolitical instability could occur (e.g. due to shifting political and economic power to East Asia and/or rising tensions in Europe), whether or not in conjunction with a new recession in the world economy. Secondly, the EU is under pressure. We see increasing differences in prosperity and rising tensions between the EU countries.

It is unclear how this will evolve. Will Europe disintegrate into different ideological regions? Or will the countries focus on themselves again? Will border controls be reintroduced? Will the Euro fall? Further extension of the EU could further undermine stability. It is also unclear what consequences Brexit will have on the EU, on the mutual relations between countries, on its economy and on the Euro.

On the other hand, climate change is a potentially disruptive force. Climate change may lead to (temporary/local) drought or flooding and raise sea levels to a greater or lesser extent, which affects the attractiveness, availability and accessibility of destinations. Climate change may also lead to (regional, national) food and water scarcity, epidemics or pandemics, which will promote a temporary or more structural shift in tourist flows and refugee flows. Especially when the growing world population is taken into account.

Altered worldview: Open worldview versus closed worldview

The (European) citizens' worldview has a major impact on their attitude to migrants and people from other countries, and eventually the mutual relationships between the countries in Europe.

In the extreme case, people may turn inwards to protect their own language and culture, and their own interests (one's own country and people), which leads to us versus them thinking and polarisation, and a short-term interest/effect that prevails over the long-term interest/effect. In that case, safety ('cure' in the short term) takes precedence over sustainability ('prevention' in the long term).

However, the other extreme is that people open themselves up to others and look for a constructive dialogue, common interests (humanity and the earth), new connections and synergy, whereby the long-term interest/effect takes precedence over the short-term interest/effect, and sustainability ('prevention' in the long term) goes for safety ('cure' in the short term).

The two key uncertainties with their extremes constitute a framework for four explorative scenarios. These alternative scenarios complement the baseline scenario. With the combination of these scenarios, it is not only the playing field in which inbound tourism could develop that can be foreseen (as defined in the phase of domain analysis), it also provides the context to prepare for such futures.

Each scenario is given a title that characterises its content, as follows:

- Global village: far away, yet close.
- Hostility: enmity towards other countries and cultures.
- Resistance: resistance to environmental disruption.
- Resilience: resilience during environmental disruption.

The alternative scenarios presented below should not be viewed as predictions, but paint four hypothetical and provocative pictures of

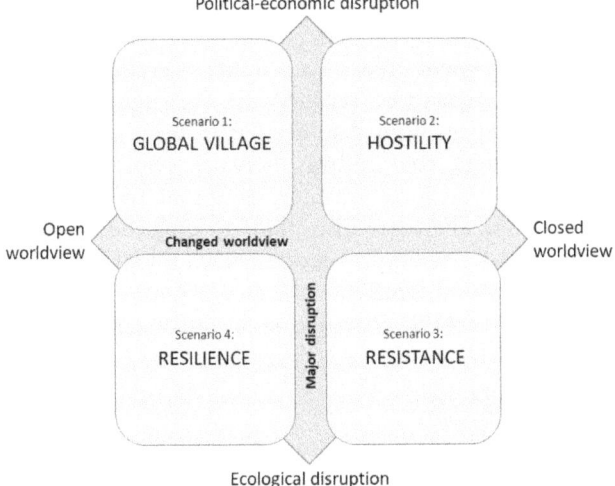

Figure 10.3 Scenario cross

the way in which the external force field that drives inbound tourism could develop. Each one is plausible and each one implies its own set of opportunities and threats. Although the expected future is *more* probable than any of the alternative futures, it is not probable in itself because anything could happen in the meantime. While each alternative future is in itself less likely than the baseline future, the alternative scenarios as a group are more likely than the baseline scenario.

10.3.1.7 Scenario 1: Global village
Open world view – political-economic disruption

It is 2030. Political and economic power have shifted to Asia, especially China. The EU has come to realise that, together, the countries can stand more effectively against the new power in the east. There is political stability, both within the EU and on the global stage.

Common (international/European) interest takes precedence over national interest. Both within the EU and between the EU and the new Asian hemisphere, there is mutual trust, synergy is sought, there is constructive dialogue and new political and economic connections and partnerships are being created.

There are no barriers to trade and international travel. The Euro is strong. Due to the growing prosperity and wellbeing worldwide, people travel extensively internationally, people learn about each other's culture and mutual understanding is fostered, which in turn benefits international cooperation. International travel has never been easier and is facilitated by mild visa procedures. Due to the absence of obstacles to international travel, sustainable (travel) behaviour is mainly based on consumers' own choices. There is a high degree of hospitality towards guests from other countries.

Substantial investments have been made over the past 10 years in the realisation of a high-speed rail network that needs to compete with flying over short and medium distances. Partly due to the contribution of Chinese technology, that railway network could be finished. It is a technological *tour de force* (masterpiece) that offers a fully fledged alternative to flying. There are also plans to upgrade train connections between Europe and Asia, using modern technology. This will further increase connectivity over land.

There are far-reaching international agreements concerning the environment.

10.3.1.8 Scenario 2: Hostility
Closed world view – political-economic disruption

It is 2030. Political and economic power has shifted to Asia, especially China. The EU disintegrated after Brexit. The situation creates political instability and fuels the threat of terrorism and war.

National interest takes precedence over joint (international/European) interest. The countries are strongly inward-looking. Both within the EU and on the global stage, there is mutual mistrust and an us versus them thinking. China is expanding its power in Europe and playing EU countries against each other.

There is fierce competition between countries on the world trade stage. Due to the protectionist stance, a trade war is still raging, resulting in all kinds of trade barriers. China is expanding its power elsewhere and seizing more and more European companies. The world economy, like the Euro, is not in good shape and a global recession is looming again. Interest in international travel has declined due to increased mistrust of other countries, increased international insecurity and complex visa and border controls. In principle, the choice to travel sustainably is left to the consumer, but when it comes to international travel, they usually deny this. Hospitality towards guests from other countries is low, even hostile.

Because travel has become more complex and time-consuming, people are looking for accommodation and entertainment more than before within their own national borders or in countries where it is still relatively easy to travel to due to individual alliances/trade agreements with trusted destinations, etc. Investments in the transport infrastructure is mainly aimed at the increased demand for domestic transport, such as, in the Netherlands, a light rail in and around the Randstad and improved train connections between the different parts of the country. This had resulted in increased domestic connectivity.

In general, countries perceive concern about the environment as a luxury problem: this is only an issue if they pose a direct threat to the economy and security.

10.3.1.9 Scenario 3: Resistance
Closed world view – ecological disruption

It is 2030. The shift in economic power to the East has been limited. The 'old' economies in the western world continue to play a significant role, but the EU is on the brink of death. It consists of different ideological zones that are increasingly drifting apart. This is partly caused by the unequal vulnerability to and consequences of climate change. This creates political instability in the EU.

National interest takes precedence over collective interest (international/European, humanity and earth). There is mutual mistrust among the dispersing ideological parts of the EU, and an us versus them thinking. While there is a lot of cooperation within regions, between regions cooperation is lacking. The fight against the consequences of climate change for water and food supplies, nature and landscape and for public health is ineffective and slow due to this lack of cooperation.

The global economy is in a trough. A neoliberalist wind is still blowing through the world, but the disappointing results in the agreements between the US and Korea, between the US and the EU, etc. at the end of the 2010s have caused sentiment to tilt. This is partly due to the fact that prosperity and wellbeing in some countries are still developing favourably at the expense of other countries. East and West and North and South are increasingly opposed to each other. At a time when climate change is unbalancing the availability of water and food worldwide, trade in these resources is vital for many countries and the risk of the spread of diseases is increasing. However, the growing polarisation means that international trade and international travel are meeting more and more resistance and that countries are not very hospitable to travellers from other countries. There is, however, a lively trade in emission standards between countries and between companies, which means that some of them remain out of the picture in the (national) approach to climate problems. Viewed on a global scale, this approach is, therefore, not very effective.

The realisation of the goals in the climate agreement is not very effective. For international long-distance transport of goods and passengers, there are major differences between the countries in laws and regulations and control and tax systems, such as CO_2 tax, tolls, 'border tax', speed traps, etc. Both discouraging the use of environmentally unfriendly forms of transport and promoting environmentally friendly slow transport depends on national coalitions that regularly change their composition. For example, some countries have chosen to concentrate their flight activities at one major airport and to phase out regional airports. Travel has become considerably more expensive. Although sustainable travel behaviour is enforced by the measures that countries take, the differences per country are large.

10.3.1.10 Scenario 4: Resilience
Open world view – ecological disruption

It is 2030. Economic power shift to the Eastern hemisphere has been relative and the 'old' economies continue to play a significant role. The EU has come to realise that the EU countries can only together come to decisive solutions for adaptation and mitigation to deal with climate change. There is political stability, both within the EU and on the global stage.

Common interest (international/European, humanity and earth) takes precedence over national interest. Both countries within the EU and beyond on the global stage are working hard to avert the threat of the effects of climate change on water and food supplies, nature and landscape and the threats to public health. There is mutual trust.

In principle, there are no barriers to trade and international travel. Visa arrangements are flexible, but the costs of travel have increased. The Euro is strong. Due to the growing prosperity and wellbeing worldwide, and the open and hospitable attitude towards guests from other countries and cultures, there is a lot of international travel, and mutual understanding is fostered, which in turn benefits international cooperation. Internationally, with the support of the IMF and the World Bank, joint stimulation programs are being set up for technological innovation, for example, in the field of world food and water production (gene technology, artificial food, 'production' of water), disease control, environmentally friendly energy production and sustainable transport systems. Food and water supply is regarded and approached as a global problem that can only be tackled by countries collectively. Sub-Saharan Africa is involved in the production of solar power. The open attitude to learn from each other creates new forms of conference and special interest tourism.

International cooperation is underway to achieve the goals in the climate agreement. International measures and standards have been agreed that are well observed by everyone. Think of regulation and sanctions on pollutive long-distance transport: CO_2 tax, toll, etc. Sustainable travel behaviour is in fact enforced through these measures.

Worldwide, international airlines and the use of airports/hubs are being reconsidered and phased out. A limited number of centrally located, climate-proof airports have been chosen with supply routes via innovative environmentally friendly forms of transport (such as, the hyperloop) that are interconnected with ultramodern environmentally friendly and climate-proof transferiums.

Due to the international attention, public opinion is turning against international travel by air. Flying is falling out of favour. Trains are hot. The use of accommodation is also changing due to the growing awareness.

10.4 Concluding Remarks

The outcomes of the study were presented at the website of NBTC Holland Marketing in an interactive manner and accessible to the public.[1] By means of a forecasting model, the NBTC used the baseline scenario to make a prognosis of inbound overnight stay tourism until 2030. This prognosis is sharpened with the help of the qualitative information from the alternative scenarios about demand, supply and environment. This resulted in a quantification of inbound tourism until 2030, including a specification by country of origin. Via these prognoses, the scenario study was used to inform the new perspective of inbound tourism to the Netherlands: Destination The Netherlands – Perspective 2030 (NBTC Holland Marketing, 2019[2]), which includes a national vision, national strategies and conditions for success for inbound tourism to the Netherlands.

10.5 Discussion Questions

(1) To what extent do you rely on quantitative forecasts when you are making plans?
(2) Can you recall any forecast(s) that did not come true? Which one(s)? Do you know what caused the discontinuity?
(3) How would you label the present era of international tourism to your destination? As the era of ….. Why? What was the disruption that made it start?
(4) How do you think international tourism to your destination will develop during the present era? Why?
(5) How would international tourism to your destination look like after 10 years?

Notes

(1) https://nbtcmagazine.maglr.com/toekomstscenario-s-inkomend-toerisme-2030-def/titelblad.
(2) https://www.nbtc.nl/en/site/destination-netherlands/perspective-destination-netherlands-2030.htm.

11 Notting Hill Carnival Futures 2020

Learning Points

- The application of scenario planning to a public mega event was designed as a bottom-up, participatory scenario planning process.
- Trust and confidence among participants are key.
- Principles from Appreciative Inquiry were applied to involve participants and gain trust.
- The participatory process was formally evaluated.

11.1 Background

The roots of the Notting Hill Carnival are in Trinidad (Google Arts and Culture, n.d.). It started in 1964 with a British social worker who decided to add a steel band procession to the Notting Hill Fayre, a tradition of the Trinidad carnival, to give expression to the Trinidadian culture. This sparked the Caribbean community who had settled in Notting Hill to spontaneously dance and jump through the streets. The first Notting Hill Carnival was held in 1964 (Ferdinand & Williams, 2018) and was visited by about 500 people (My London News, 2015).

Over the years, the Carnival has evolved from a fair extended with a steel band procession to an internationally acknowledged multicultural, mega-event with food, drink, music, dance and masquerade. The Notting Hill Carnival has grown into a multi-cultural event, with Trinidadian, Jamaican, Brazilian, African and British influences and is held every August Bank Holiday weekend.

It has expanded into one of the biggest street festivals in Europe, with an estimated amount of 1.5 to 2 million visitors during the three-day event. It draws an international audience with 44% coming from outside London and is largely carried by volunteers. It is regarded to have significant positive social and economic impacts and has been the source of inspiration to other carnivals in the UK and around the world (such as, in Rotterdam and Berlin). Moreover, Notting Hill carnivalists are in demand worldwide for appearances at other carnivals and cultural events, both throughout Europe and elsewhere (such as, in Korea, Japan and Nigeria).

Figure 11.1 Impression of crowds at Notting Hill Carnival
Photo by Bernd Dittrich on Unsplash.

However, as it approached its 50th anniversary in 2014, the carnival had been facing tough challenges in the field of public and private funding, overcrowding in the narrow streets (Figure 11.1), racial prejudices and racial politics, public disorder and crime, along with an ageing and declining number of the festival's participants. This led to concerns about the preservation of the carnival's traditional arts forms, crowd management, public safety and surveillance. The Carnival struggled with the external image among popular media and academics that cultural organisations had a lack of entrepreneurial abilities and that trustees, organisers and administrators were incompetent and corrupt (Ferdinand, 2013).

The 50th anniversary of the Notting Hill Carnival in 2014 would be a great moment for reflection, to look to the future and explore a new vison and strategic lines. A range of Carnival partners took the initiative to collectively consider the challenges, opportunities, uncertainties and future plans by means of a scenario study entitled *Carnival Futures: Notting Hill Carnival 2020*. King's College London invited the European Tourism Futures Institute (ETFI) to guide a bottom-up and collaborative scenario planning process with as many stakeholders involved as possible. The project would bring together the festival's organisers, attendees, funders and other supporters. The Carnival community, the High Commission for the Republic of Trinidad and Tobago and Caribbean Enterprise Network UK (CENUK) supported the initiative. A dedicated project logo was designed (see Figure 11.2) and a website was created to communicate information and newsletters about the project.[1] The project was commissioned and funded by the King's Cultural Institute's Creative Funding programme and formally evaluated by Bournemouth University afterwards.

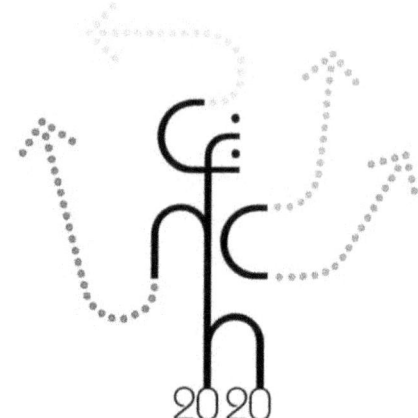

Figure 11.2 The logo of the Notting Hill Carnival Futures 2020 project

11.2 Purpose

The scenario project Notting Hill Carnival Futures was conducted in 2013. For the ETFI it was its first international project. The main purpose of the project was to engage the Notting Hill Carnival Community in collaboratively planning for its future together with all stakeholders, academics and other relevant individuals who could contribute meaningfully to the planning process. The fact that the previous strategic review of the Carnival was done in 2004 and the 50th anniversary of the Carnival was upcoming (in 2014), the project was taken as an opportunity to collectively consider the challenges, opportunities and uncertainties that the carnival was facing and to plan for the future. It also sought to develop new and more meaningful partnerships between academia and the cultural community. Another aim was to provide inspiration for the cultural and entrepreneurial innovations which would sustain the Notting Hill Carnival in 2020 and beyond.

11.3 Approach

Scenario studies, such as the one that was conducted for the Notting Hill Carnival, can be positioned within the emerging approach of strategic foresight (see Chapter 4). Scenarios take the key uncertainties that drive the future of organisations as a starting point. Combinations of the possible directions to which such uncertainties could develop are used to frame a number of scenarios as 'painted pictures' of uncertain yet plausible futures and not as predictions of the future. Not one of these scenarios is more likely than the others – they are just alternative pictures of what could happen. The lively explorations offer inspiration for decision makers to foresee and anticipate future changes proactively

Figure 11.3 Impression of performance at Notting Hill Carnival
© Euphoria Carnival.

by means of new concepts, business models or strategic courses of action. The process to arrive at these scenarios is at least as important as the outcome! Ideally, both the key uncertainties – the framework on which the scenarios are built – and the contents of the scenarios are constructed from the perception and understanding of the participants involved.

The project kicked off with a presentation to representatives of the key stakeholder groups: the Carnival community, the High Commission for the Republic of Trinidad and Tobago and the Caribbean Enterprise Network (CENUK.) The scenario planning project took a stepwise, bottom-up approach while some of the principles of Appreciative Enquiry were integrated into the process. Rather than acting as experts, the scenario planners of the ETFI worked as facilitators, allowing the stakeholders of the Notting Hill Carnival to lead the research process themselves. During three subsequent half-day workshops, carnivalists, public and strategic experts holistically elaborated their views on the core values of the carnival, its unique selling points and its challenges and future uncertainties. The time horizon for the scenarios was 2020.

11.3.1 Notting Hill Carnival organisations: Core values of the Carnival

The first workshop (held on 23 September 2013) was with the carnival organisations who are responsible for the organisation and the staging of the Notting Hill Carnival: steel bands, dancers, calypsonians,[2]

Figure 11.4 Impression of performance at Notting Hill Carnival
© Panpodium.

costume makers and organisations playing recorded music to source the street parties and entertaining large groups of young people who come to the festival: the British Academy of Sound Systems, Disco Hustlers, Elimu Paddington Arts Mas³Band, Genesis Mas Band, Mangrove Steelband, Muzik Lil Muzic, Masqueface, Nostalgia Steelband, Reading AllSteel Percussion Orchestra and Rough but Sweet Sound System (illustration of the Notting Hill Carnival in Figures 11.3, 11.4. 11.7, 11.15, 11.16, 11.17, 11.19). The aim of this workshop was to identify the core values of the Carnival for the community.

For a successful workshop, in which the participants show an open attitude and are willing to sharing experiences, it is important to have trust and confidence in each other and in the facilitators (white, from another country and academia). Given the negativity around the Carnival, and to prevent a cynical attitude of the carnivalists, during the workshop the philosophy of Appreciative Enquiry was used to create a positive atmosphere. Instead of focusing on fixing weaknesses, a common approach in strategic planning and change management, Appreciative Inquiry is an approach that emphasises inquiry of strengths. It seeks to uncover what is going well, to understand why it works well and to explore how these positives can be amplified and transferred to other areas. It is an approach in which the participants feel heard and supported because they contribute to a positive future state. Most commonly, Appreciative Inquiry is subdivided into five phases: define (decide upon what, who and how), discover (appreciating the best of what is), dream (envisioning of what could be), design (co-constructing of what should be) and deliver (co-constructing of what should be (sustaining of what will be) (Cooperrider & Whitney, 2005).

The participants were grouped randomly at a number of tables. Key points on the agenda were as follows:

(1) Individual exercise: participants were asked to recall a meaningful or precious positive event, situation or process that they had experienced during the Notting Hill Carnival that:
    - made the participant most proud;
    - enabled the participant to feel most involved;
    - the participant found most funny, nice or motivating;
    - the participant valued most; and
    - then describe that memory on a proforma that was distributed by the facilitators.
(2) Group exercise: participants at a table share their stories with each other (discuss it, ask each other questions, etc. See Figure 11.5).
(3) Individual exercise: participants were asked to underline or highlight one or two features of the experience that they had written up that made it so valuable to the participant. These were 'core values' of the event and were written on separate hexagons.
(4) Group exercise: clustering of the hexagons on the wall to create a thickened list.

Figure 11.5 Participants from the Notting Hill Carnival community sharing their experiences
Photo by Albert Postma.

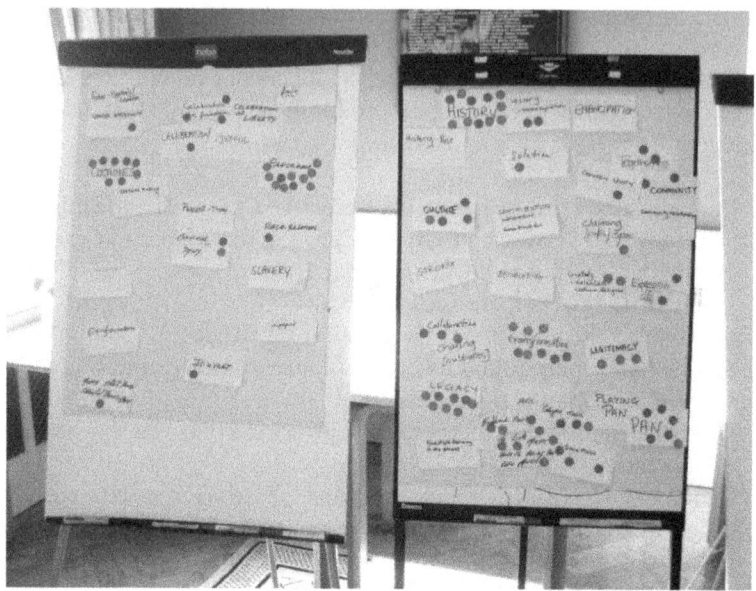

Figure 11.6 Core values with dot-votes representing appreciation (green stickers) and uncertainty/doubt (red stickers) concerning the future of the Notting Hill Carnival

(5) By means of a voting process with red and green stickers, the participants were asked to identify the most appreciated (green sticker) and the most doubtful/uncertain core values (red sticker) for the future of the Carnival (see Figure 11.6). Through this process consensus was achieved which produced the list in Table 11.1, in which the number reflects the level of appreciation and the level of uncertainty and doubt ascribed to each item.

Table 11.1 Core values of the Notting Hill carnival, as perceived by the participants

	Level of importance	Level of uncertainty
Music	9	5
Costumes	6	2
Pan	5	1
Community	5	0
History	4	7
Celebration of freedom	4	0
Legacy	3	5
Panorama	2	9
Transformation	2	6
Legitimacy	0	3
Creativity	0	2

Figure 11.7 Impression of performance at Notting Hill Carnival
© John Lastaukas.

Carnivalists identified music as the most important core value of the Notting Hill Carnival. The core values that they were most worried about were Panorama (the steel pan event held just prior to the Notting Hill Carnival bank holiday), the historic value and transformation.

(6) **Group exercise:** at each table the participants were asked to discuss the possible consequences for the Carnival if the core values would either be ignored or nurtured. These were collected on hexagons and put on the wall, at opposite poles, at both sides of the core values – such as illustrated in Figure 11.8.

11.3.1.1 Notting Hill Carnival's public: Unique selling points of the Carnival

The second workshop (held on 24 September 2013) was aimed at the following, partly overlapping, groups (Figure 11.9):

- Visitors to the Notting Hill Carnival.
- Repeat visitors (those that attended Carnival several times over many years).
- Londoners, particularly Notting Hill residents, who have grown up with the Carnival.
- Members of the Carnival Diaspora – those who followed the international circuit of Trinidad-style or Caribbean carnivals worldwide.
- People who had stopped coming and wanted to come again but hadn't attended yet for different reasons.

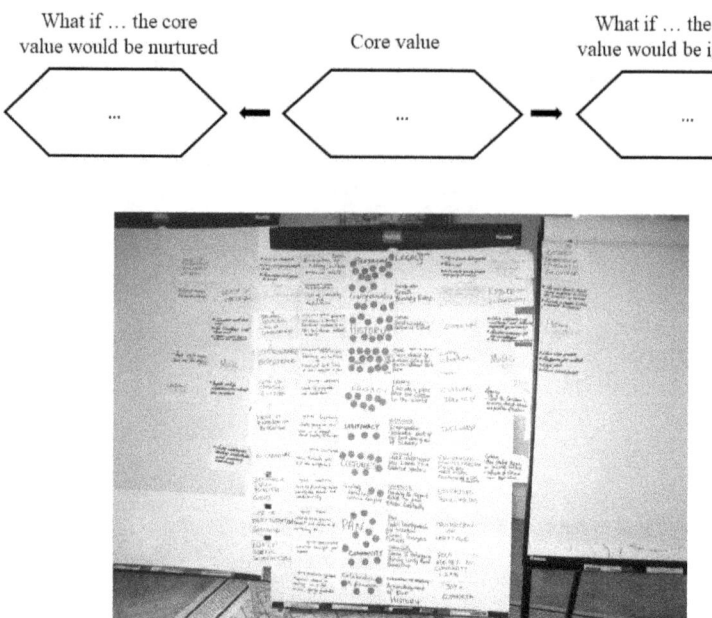

Figure 11.8 Perceived implications if the core values would be nurtured or ignored

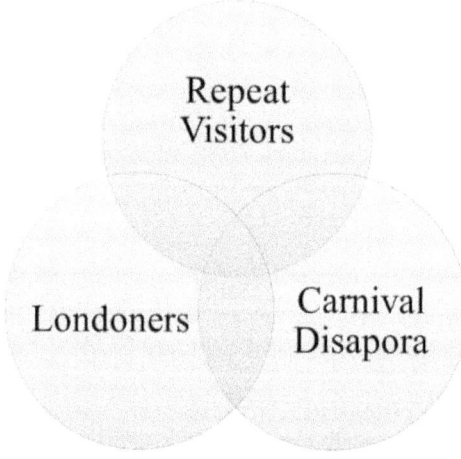

Figure 11.9 Composition of Notting Hill's Carnival Public

There are often times where there is a mismatch about the organisers' and the visitors' views regarding what the festival represents. The intention of this workshop was to capture the visitors' viewpoints and to recognise the Carnival's unique selling points. The structure of the

Figure 11.10 Core values with dot votes for importance (green/lighter stickers) and worries about the future (red/darker stickers) by visitors of the Notting Hill Carnival

workshop was similar to the one with the Carnival organisations. The participants were asked to recall a precious or valuable moment at the Notting Hill Carnival they had visited. Participants who had yet to visit the Carnival were asked to imagine the Carnival as they would like it to be. The participating members of the public had to list three core values in their story. When the lists were combined, each of the participants ranked the core values according to their perceived importance and their perceived worry about the durability of that core value (see Figure 11.10). This process resulted in a list of core values (see Table 11.2), to be interpreted as unique selling points.

The public perceived tradition as the most important core value, or unique selling point, followed by the freedom of spirits in the street and sense of unity. They felt most worried about the future of the freedom of spirits in the street and community.

Like the carnival organisations, the public was asked to discuss in their groups the possible consequences for the Carnival, if the core values or unique selling points (USPs) would be either ignored or nurtured (Figure 11.11). The output of the discussions is listed in Table 11.2.

11.3.1.2 Strategic experts: Driving forces and key uncertainties

The third and last workshop involved experts, which included public sector employees, researchers, event managers, town planners and festival network organisations, who are together responsible for the context in which the carnival is staged, currently and in the future (see

Table 11.2 Unique Selling Points of the Notting Hill Carnival as perceived by a group of visitors

	Level of importance	Level of uncertainty/worry
Tradition	11	1
The freedom of spirits, experienced in the streets	8	7
Sense of unity	8	1
Safety	7	4
Freedom	7	2
Mas' (masquerade)	6	3
Youth involvement	6	1
Uniqueness	6	0
Artistic aspects	4	2
Pan	4	1
Street	4	1
Community	3	7
Culture fusion	3	5
Cultural experience	3	1
Culture	3	1
Family	3	0

Figure 11.12): Backstage Boutique, Bahama Host Association, Brent Council, the Department of Health, the Luton Hat Factory, SABMiller plc, Shine Live, Trinidad and Tobago High Commission in London, Trinidad and Tobago Tourist Board and Vintage Misfits, Anglia Ruskin University, Arts Council England, Cardiff Metropolitan University, Crewsaders, the Caribbean Enterprise Network, Goldsmiths, University of London, International Special Event Society (ISES), Royal Borough of Kensington and Chelsea Council and the University of Warwick. The aim of this workshop was to develop the framework for the four scenarios.

Figure 11.11 Two USPs identified by the participating visitors

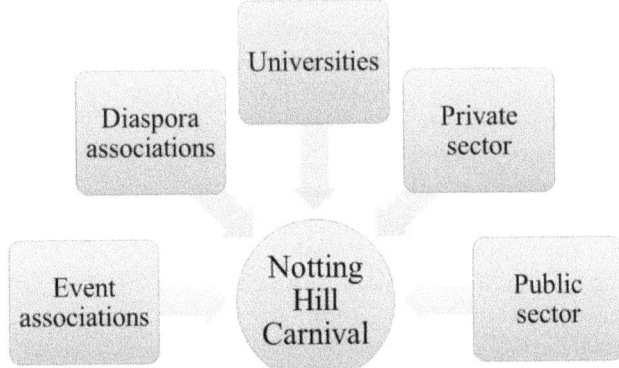

Figure 11.12 Festival network organisations

The experts worked in groups of three or four. Each group was asked to identify the most important/powerful driving forces that would affect the future of the Notting Hill Carnival, together with the possible extremes to which the driving forces might lead until 2020. The participants were challenged to consider demographic, economic, social/ cultural, technological, ecological and political/legal/institutional drivers (see Figure 11.13). Eventually, they identified the following

Figure 11.13 Experts of festival network organisations at work

external developments to be important for the future of the Notting Hill Carnival:

(1) Increased house prices and gentrification in Notting Hill.
(2) Increase in security, health and safety issues.
(3) Political change and decreased funding.
(4) Change in London's ethnic make up.
(5) Differentiation of attendee groups.
(6) Legal restrictions concerning noise, pollution, etc.
(7) Changing identity of London as a world city.
(8) Notting Hill Carnival's location.
(9) Political developments affecting residents and businesses.
(10) An ageing West-Indian community and the dying out of the 'Windrush generation'.
(11) Support from politicians and a cohesive community.
(12) Emerging new generation of 'makers and producers'.
(13) Cultural diversity.

Once the driving forces were collected, the participants were asked to rank them all according to level of importance and level of uncertainty by means of green and red stickers, respectively (see Figure 11.13). The driving forces that were both the most powerful and at the same time the most uncertain were identified as the two key uncertainties to form the two axes of the scenario cross.

From this analysis it was clear that two driving forces stood out from the rest: political change resulting in decreased funding (a key uncertainty), and the emerging new generation of 'makers and producers'. These two factors have been labelled in the scenario cross as the 'financing model' and the 'community composition'.

11.3.1.3 The four scenarios

The viewpoints of the experts (workshops 3) were used to establish the scenario framework defined by two key uncertainties for the period from 2014 to 2020: the financial model and the community composition of Notting Hill. The key points that were brought forward by the participants of the other two workshops had been used as input for the contents of the four scenarios. The opposites of what could happen when the core values or the USPs would be nurtured or ignored was woven into the narratives of the most appropriate scenario quadrant. Eventually, the scenarios explore four different pictures of how the Notting Hill Carnival might have evolved by 2020, when the two most influential uncertainties develop in different directions. The scenarios should not be regarded as predictions of the future but as positive descriptions of the Notting Hill Carnival's possible futures that can happen (Figure 11.14).

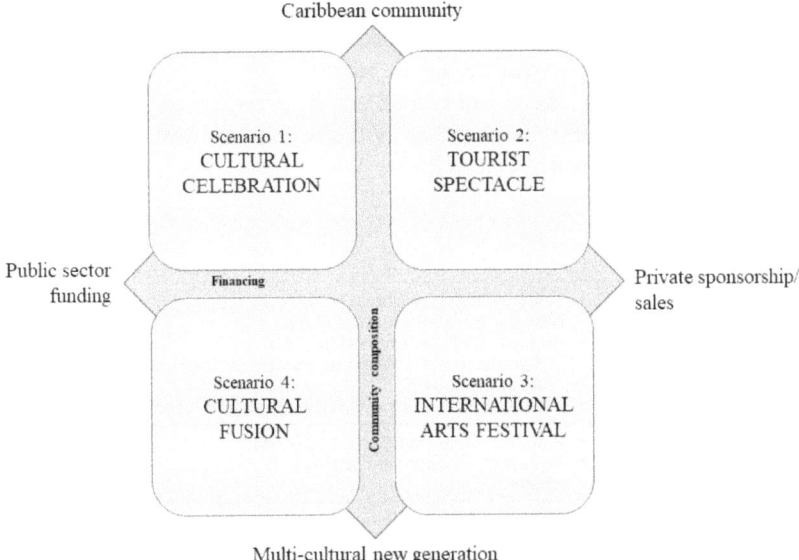

Figure 11.14 Four scenarios for Notting Hill Carnival Futures 2020

11.3.1.4 Scenario 1: Global village
The Caribbean community – public sector funding

In this scenario, the Notting Hill Carnival would be seen as a way for the remaining Caribbean community in the UK to express its identity. They would celebrate their cultural history with traditional Trinidadian calypso and Soul Calypso (soca) music. Live steel bands and traditional costume making would be the central focus of this event. Non-Caribbean people would be free to attend and join in the activities, however, their presence would be secondary. This would be a smaller, more focused carnival for and by the Caribbean community.

There will be a notable absence of sponsorship banners and logos at the event and its principal source of funding will come from UK arts funding bodies and the third sector, with some assistance from Caribbean businesses and associations. The organisers and performers will not be commercial employees but committed volunteers from families with a long history of serving or participating in the Notting Hill Carnival.

There will also more people attending with children and as families. School teachers and university lecturers will view this event as a place where they can have their students learn about the history and culture of the Caribbean community in the UK. Historians and researchers will be drawn to the festival as well. Additional events, such as Caribbean film screenings, theatre performances and art exhibitions will be added to heighten the festival's educational aspects. There will also be workshops dedicated to passing on traditional art forms.

This is not an event concerned with drawing huge crowds or spectators. Many of the attendees will also be cultural practitioners, participating in the festival as well.

Future points

Nature of event	• Celebration of Caribbean cultural history: traditional music, live steel bands and costume making.
	• Additional educational events, such as:
	  – Caribbean film screenings, theatre performances and art exhibitions; and
	  – workshops dedicated to passing on traditional art forms.

Spin-off	• Preservation of traditional arts forms. • Expression of Caribbean identity. • Education.
Size of event	• Small and focused carnival of two days.
Carnival organisation/ performers	• Caribbean community. • Organisers and performers are committed volunteers. • Non-Caribbean people join in the activities.
Funding	• Public funding. • No admission fee. • Primary funding from public funding from UK arts funding bodies and third sector. • Secondary funding from Caribbean businesses and associations.
Audience	• Caribbean community, with more young families and children. • Students, historians, researchers that want to learn about history and culture of the Caribbean community.
Location	• Notting Hill.

11.3.1.5 Scenario 2: Tourist spectacle
Caribbean community – private sponsorship and sales

In past years, the Notting Hill Carnival has had to deal with decreased funding by both the public and private sector and increased restrictions by local security, health and safety regulations. In this scenario, to be able to survive, the Carnival organisations will seek to create a more controlled and safer environment, which uses alternative sources of funding. They will utilise new innovative methods, such as text donations and crowd funding (where individuals will collectively pool their money via the internet to sponsor the carnival), to raise revenues. The spatial size of the festival will be limited and it will be moved to an enclosed venue. Attendees will pay an entrance fee. The Notting Hill Carnival will mimic other commercial arts festivals and have a programme of activities which will be offered over a full week. These ticketed events will feature a varied programme of performances. Much like other top commercial festivals, only the best acts will be

Figure 11.15 Impression of performance at Notting Hill Carnival
© Association of British Calypsonians.

selected to perform at this carnival. The intent being that the high quality of talent will draw international tourists and sponsorship from large organisations – which would be necessary, as competitive fees would have to be paid to consistently draw artists who are regarded as the best in their respective genres.

However, in this scenario the organisers will have to retain the authentic carnival spirit, as it is a crucial USP. All the traditional art forms will be preserved, such as soca, calypso, pan and costume making. They will be featured alongside more commercial acts, such as top DJs, hip-hop, grime and reggae artists. The festival will be a more mainstream and commercial presentation of the Caribbean culture which will broaden its appeal. The focus of this event will be the attendees.

Future points

Nature of event	• Mainstream commercial arts festival with varied programme of (expensive) top performances in varied genres. • Caribbean culture is presented in a commercial way. • Authentic carnival spirit with traditional art forms in conjunction with new and commercial art forms. • Focus on attendees.
Spin-off	• Preservation of traditional arts forms and authentic carnival spirit. • Broadened appeal.
Size of event	• Limited spatial size, duration of a week.
Carnival organisation/performers	• Commercial event organisations.
Funding	• Entrance fee (ticketed event). • Alternative innovative sources of funding (text donations, crowd funding, sponsorship, etc.).
Audience	• Large number of international tourists.
Location	• Controlled and safe environment, enclosed venue.

11.3.1.6 Scenario 3: International arts festival
Multi-cultural new generation – private sponsorship and sales

The Notting Hill Carnival has proven to be a major event that attracts millions of visitors with a significant economic spin-off for London. In this scenario, London's tourism authorities will become more

Figure 11.16 Impression of Notting Hill Carnival
© Eduardo Noriega, London School of Samba

involved in the organisation and commercialisation of the event. To achieve commercial success, the authorities will transform the Carnival into a truly London event with a greater representation of all London's communities. The festival will have a broad appeal.

The Notting Hill Carnival will grow into an internationally recognised 'arts festival' with the inclusion of many ethnic groups that reflect London's diverse ethnic make-up. At the festival, there will be a significant place for the cultural arts forms of the Caribbean community, but they will be presented as artistic performances rather than celebrations of cultural history. Best practices in events management will be incorporated into the festival, such as sustainability and the adoption of cutting-edge technologies. The Carnival will take place in venues across London.

The incorporating of London's diverse ethnic communities will result in international participants and attendees being drawn to the festival. The Notting Hill Carnival will be a world Mecca for carnival arts and art forms. It will utilise London's position as a world city to its full advantage, resulting in international media attention. Broadcasts from the festival will be seen in many countries across the globe. Key sources of funding will be from the sale of broadcast rights and sponsorship from multi-national corporations. This will be a festival intended for a global audience.

Future points

Nature of event	• Internationally acknowledged multi-ethnic arts festival. • Significant place for cultural arts and arts forms of Caribbean community. • Commercial and touristic London-event emphasising London as a world city. • Artistic performances rather than celebration of cultural history.
Spin-off	• Multi-ethnic representation of London. • Significant economic spin-off for London. • International media attention (broadcasts around the world).
Size of event	• World mega-event with broad appeal.
Carnival organisation/performers	• Organisation and commercialisation of event driven by London's tourism authorities. • Best practices in events management (sustainability, cutting edge technologies, etc.).
Funding	• Sale of broadcast rights. • Sponsorship from multinational corporations.
Audience	• Global audience. • International participants and attendees.
Location	• Venues across London.

11.3.1.7 Scenario 4: Cultural fusion
Multi-cultural new generation – public sector funding

The population of Notting Hill has gradually been changing because of the gentrification processes. Housing prices have increased and houses that were left by members of the immigrant Caribbean community are now occupied by young and affluent families, typically not of Caribbean origin.

In this scenario, the population change will be exploited as an opportunity to feed the Notting Hill Carnival with the inspiration of a new generation of 'makers and producers' that break with the conventional rules and traditions. This carnival will offer a safer environment for families with children and more room for family friendly experiences.

The festival will expand into a large green open space, like many other family oriented events. This expansive space will allow for more freedom of expression and, thus, new forms of artistic expression will be stimulated which will attract additional public sector funding, further encouraging the creation of fusion art forms.

The involvement of the new citizens will result in a diverse yet harmonious expression of music, art and culture. The Carnival will grow into a unique fusion event with increased levels of participation, empowerment and creativity. Participants and attendees will appreciate and learn from each other. There will be no social tension between the old and new residents. Community ownership of the event will be revived and the role of grass roots organisations and volunteers will increase dramatically. In fact, the Carnival will evolve into a showcase of new hybridised global cultures.

Future points

Nature of event	• Unique fusion event that adapts to changing community composition in Notting Hill and breaks with conventional rules and traditions. • More freedom of expression; high level of creativity; new forms of artistic expression; fusion arts forms. • Diverse yet harmonious expression of music, art and culture.
Spin-off	• Showcase of new hybridised global cultures. • Appreciation and learning between old and new residents of Notting Hill. • Revival of carnival ownership.
Size of event	• National event.
Carnival organisation/performers	• Prominent role for grass roots organisations and volunteers. • High level of participation and empowerment.
Funding	• More public funding.
Audience	• Londoners.
Location	• Safe environment for families with young children.

11.4 Concluding Remarks

The Notting Carnival Futures 2020 project was intended as a contribution to current research and to identifying future directions for the development of the Notting Hill Carnival. The ETFI designed a tailor-made scenario planning process that suited the needs of

Figure 11.17 Impression of Notting Hill Carnival
Photo by VENUS MAJOR on Unsplash.

the Carnival Futures: Notting Hill 2020 project. Through a genuine collaboration between academia and cultural consumers, organisations and policymakers, the innovative scenario-planning and foresight method yielded a suite of provocative possible futures for the Carnival with the intention to provide the inspiration for the cultural and entrepreneurial innovations which would support and sustain the Notting Hill Carnival in 2020 and beyond.

One of the key benefits of the project was the involvement and engagement of such a broad variety of stakeholders and the positive, open and constructive way in which Notting Hill Carnival's future was explored collaboratively. Findings from the project were disseminated in a variety of ways. The project resulted in Carnival Futures short films containing excerpts from participant workshops, the Carnival Futures digital monthly newsletter containing the key messages and other updates about the project, the Notting Hill Carnival 2020 booklet with the key findings targeting ordinary members of the public (the ETFI report), a Notting Hill Carnival 2020 stakeholder report aimed at those organisations which assisted in creating the research, a Carnival Futures white paper targeted at government agencies involved in the Notting Hill Carnival, a Carnival futures journal publication, a Notting Hill Carnival 2020 launch event and a documentary of the three workshops produced by The New Media Angels.[4]

It should be noted that the scenarios are not the end of the process, but just the beginning. They should form the input for further discussion in which the implications for each of the scenarios are mapped in detail, valued, clustered and translated into courses of action. Even though the carnival's organisers may wish to work toward a particular scenario, to make the Notting Hill Carnival future proof they should be prepared for all of the four scenarios and not choose only one that seems most preferable. So, ideally, the courses of action they should take for 2020 should transcend the individual scenarios. The key areas highlighted by the scenarios for reconfiguration are the event's cultural aspects, its size, its human resources, the funding model, promotion/branding and the location. In the festival, as it was in 2013, aspects were present that could be developed to fashion each of the scenarios presented.

The constraining of the festival's size to match its level of resources will be critical. For three of the four scenarios, alternative sites for the carnival to be hosted also need to be considered. According to the scenarios, enhancing the human resource capacity, funding sources and promotion/or branding of the carnival is necessary for any of the scenarios to be successful.

Three key recommendations were formulated at the end of the project:

- To develop Notting Hill Carnival's human resource capacity by the engaging of younger people (whatever their origin) is absolutely essential because the older generations of traditional carnivalists are dying out. This requires the current carnivalists to embark on a deliberate strategy of transferring of knowledge, skills and also leadership responsibilities to younger people.
- The Notting Hill Carnival has a great deal to offer both artistically and commercially. In the current financial climate, there needs to be a greater level of innovation applied in exploiting these aspects. The international network that the Notting Hill Carnival belongs to can also be explored since the local funding environment is so competitive.
- The development of a distinctive brand identity, which is in line with both the attendees and funding bodies requirements, is also essential if the Notting Hill Carnival is to develop a sustainable business model in the long term.

It was believed that by developing these core areas, the organisers would have the greatest potential for developing the cultural and entrepreneurial innovations needed to support the future of the Notting

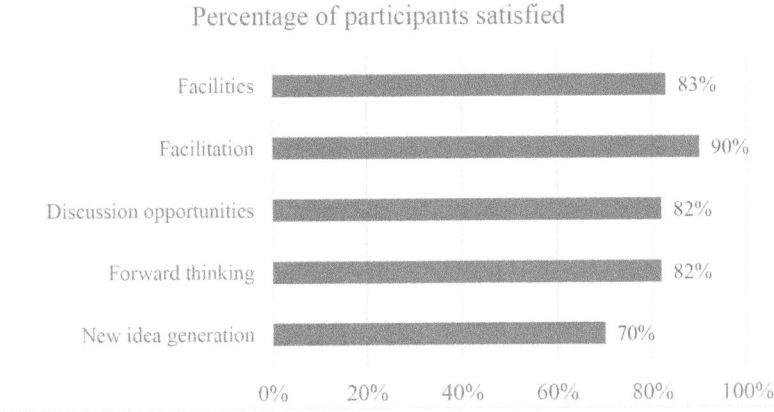

Figure 11.18 Overall participant satisfaction with workshop aspects

Hill Carnival. So, as festival organisers looked toward 2020, developing human resource capacity, funding sources (whether public or private) and the promotion/branding of the carnival should have been the key priorities.

11.4.1 Independent evaluation

An independent evaluation of Carnival Futures: Notting Hill Carnival 2020 was carried out by the School of Tourism in Bournemouth University. Forty six of the 55 distributed questionnaires were returned.

Figure 11.19 Impression of performance at Notting Hill Carnival
© Association of British Calypsonians.

The evaluation survey showed that the workshops participants came from a diverse range of ethnic backgrounds and nationalities. The majority (28 participants) identified themselves as British, with eight of these identifying as Black British or Black Caribbean British. There were also a significant number of participants who were Caribbean nationals, the most popular island being Trinidad & Tobago – some seven participants were nationals of this country. Other participants came from European countries outside the UK, with a few from other international destinations. Three declined to identify their nationalities. The participants were spread across all age groups, reaching from 16–24 to 65+. The ratings of the workshops were positive, for the vast majority of participants. For each of the five aspects the workshops evaluated by the questionnaire, at least 70% of participants reported that they were satisfied (Figure 11.18).

In the survey, participants were also asked to comment on their vision for Notting Hill Carnival 2020, which may have been sparked by the workshops. To that end, 94% of all workshop participants made positive comments in relation to their vision.

11.5 Discussion Questions

(1) Could you make a list of the key stakeholders (internal and external) of your event, business, organisation or destination including their main interests?
(2) Could you indicate of each stakeholder whether they are dominant (those with a formal connection to your organisation, such as Board of Directors or Public Relations) and/or dangerous (those with powerful and urgent claims, and are coercive and possibly violent for your organisation, such as activists or certain employees), and/or dependent (those who are dependent on your organisation to carry out their will, such as local residents or nature)?[1]
(3) Have you considered to include them in your long term planning? Why? Or why not?
(4) How would you motivate them to collaboratively think about, debate, and shape the future of your event, business, organisation or destination? (see also Chapter 4.7).
(5) What would be important aspects for you along which the quality of the participatory process could be evaluated?

Acknowledgements

Carnival Futures: Notting Hill Carnival 2020 was a King's Cultural Institute project led by doctoral student Nicole Ferdinand from the Department of Culture, Media Creative Industries at King's College London. The project was informed by her research expertise in the areas of international business theory, cultural tourism and festival making. The project was commissioned and funded by King's Cultural Institute's Creative Futures funding programme.

It was operated with the support of Association of British Calypsonians, British Association of Steel Bands/Panpodium, British Association of Sound Systems, Caribbean Music Association and Carnival Village & ABC. The project was supported by various Carnival organisations, in particular the Association of British Calypsonians (ABC), the British Association of Sound Systems (BASS) and the Caribbean Music Association (CMA), the British Association of Steelbands (BAS), Next Level + PoisonUK events and Metro Glory Music. Carnival Futures: Notting Hill Carnival 2020 was able to benefit from the free use of venues and facilities from Carnival Village and the High Commissions of both Trinidad and Tobago and St. Lucia in London.

Notes

(1) https://www.kcl.ac.uk/cultural/projects/2016/carnival-futures.
(2) Calypso is a dance music style from the Caribbean, originating from social conflict in Trinidad.
(3) Mas is short for Masquerade. It refers to the producers of the costumes that are worn in the parade.
(4) https://www.kcl.ac.uk/cultural/-/projects/carnivalfutures and https://youtu.be/ef4jAXXQu8Q.

12 Visitor Pressure in European Cities

Learning Points

- One of the most difficult things in managing visitor pressure is to get all the relevant stakeholders on board. Particularly in cities, where there is a wide variety of different perspectives, it is difficult to engage in a meaningful discussion.
- The UNWTO, and partners involved, have shown less interest in the scenario part of the visitor pressure study, and in anticipating overtourism and adaptive capacity of cities in the future, than in strategies to manage the issue right away. Therefore, a short-term focus seems to prevail.
- Scenarios have to be critically reviewed at regular intervals and need to be adapted to anticipated changes in the environment.

12.1 Background

In 2014, the European Tourism Futures Institute (ETFI) signalled various messages in the media about the irritation of residents in cities, such as Venice and Barcelona, towards tourism in their cities. Around that time, the media started to report on a negative attitude among the local population to visitors, due to issues with overcrowding, noise and other nuisances, supposedly caused by tourists. Anticipating that the reporting by the media about visitor pressure could be an early indication of an issue that might become more important over the years, the ETFI felt the urge to initiate a study on this phenomenon. The ETFI conducted a pilot study in Amsterdam, Berlin and Riga, of which the outcomes were presented at the ITB in Berlin March 2015 (Postma, 2015a). After consultation with the director of the Dutch Center of Expertise in Leisure, Tourism and Hospitality (CELTH[1]), it was decided to put the follow-up under the responsibility of CELTH. In his previous job, the director had been director of Amsterdam Marketing and, therefore, part of a big network of Destination Management Organisations of cities in Europe[2] This facilitated access to other cities, which resulted in the participation of Barcelona, Lisbon, Munich,

Amsterdam, Berlin and Copenhagen, and in the second round of Tallinn, Salzburg and five cities of art in Belgium.

Visitor pressure, in recent years mostly referred to as overtourism (Koens *et al.*, 2018b), is a complex phenomenon because it is caused by many interrelated factors, which is for a part directly associated with tourism. Visitor pressure is often referred to as the overuse of urban resources, which may be caused by sharp increase in the number of visitors and the number of festivals in cities, often with peaks in specific periods of the year, the advent of Airbnb, and, in association with the previous causes, the spreading of tourists across urban areas where residents do not expect them, in their search for authentic and undiscovered places (Koens & Postma, 2017; Wolfram & Burnill-Maier, 2013). However, is also likely that acceptance and tolerance have declined (Postma, 2013b).

Visitor pressure is indirectly caused by societal factors outside tourism. Globalisation and informatisation have made tourism interdependent on what happens across the borders of political territories, domains and industries. Many developments in the global society have implications for tourism, whether they are demographic (size and nature of the population, urbanisation), economic (i.e. emerging economies, recessions, labour issues, etc.), social (i.e. changing value patterns, customer behaviour), technological (i.e. transport technology, robotisation, virtual reality), ecological (i.e. climate change, see level rise, precipitation, drought) and political (i.e. populism, power shift to Asia, measures to fight $CO2$ emissions). Such implications will affect the size of travel flows and the distribution of tourist origins and destinations around the world, thus, stimulate or harm tourism in specific destinations. Therefore, it is important to put visitor pressure, or overtourism, in a wider urban context. For a full understanding, it is key to realise that international tourism is but one of the usages of the city space. The developments go fast and the interrelations between all these factors are volatile, ambiguous and complex and future demographic, economic, social, technological, environmental and political developments will have its impact on how city tourism and overtourism may evolve during the years to come. Strategies to manage overtourism are largely related to our present understanding of the nature and scope of this phenomenon. Such strategies should take the city context, the wider city governance structure and wider societal factors into account in order to be successful (Koens *et al.*, 2018b; UNWTO, 2018).

For strategies to be successful, they should anticipate such changes. Strategic foresight and scenario planning provide an approach to envision the changes that could occur in the longer term. Scenarios are hypothetical yet plausible futures that can be used to 'wind tunnel' test existing strategies or to develop new ones. Ideas that fit into all scenarios would be most robust and would prepare the city the best for future uncertainties (Koens & Postma, 2017).

12.2 Purpose

In this chapter, a strategic foresight and scenario study is presented that was executed as part of a Visitor Pressure project that was operated by CELTH. The aim of the project was to get a better understanding of visitor pressure/overtourism as perceived by the cities' residents (survey) and experts (interviews) and to explore the future development of visitor pressure in a European urban context by means of scenarios. The purpose of this chapter is to present the scenarios and how they were developed, plus the way the scenarios were used to evaluate contemporary strategies to manage overtourism and to review important implications of the scenarios to inform the development of new strategies.

12.3 Approach

The visitor pressure study by CELTH, and so the scenario study as part thereof, was conducted in two rounds, with a different set of clients. The first round was commissioned by the Destination Management Organisations (DMOs) of Copenhagen, Berlin, Munich, Amsterdam, Barcelona and Lisbon. The second round by the DMOs of Tallinn, Salzburg and the five Belgian cities of Art Leuven, Antwerp, Ghent, Bruges and Mechelen. During both rounds, representatives of the DMOs of the cities participated in a workshop at Amsterdam Schiphol Airport, facilitated by the ETFI. In the first round, scenarios were developed for the sustainable development of urban tourism in Western European cities (Koens & Postma, 2017). In the second round, the process was done from scratch and based on the findings the scenarios form in the first round being slightly adjusted (Postma et al., 2018). Besides, the scenarios were used to wind-tunnel test strategies that were identified by means of desk research and listed in the report. In a wind tunnel, the prototype of a new car or airplane is tested against external circumstances. In the context of scenario planning the expression 'wind tunnel' is used metaphorically, to clarify that existing or new strategies are tested against the external circumstances as they occur in each of the scenarios. A wind-tunnel test shows if and what has to be done in order to make the strategies future proof.

12.3.1 Round 1: Copenhagen, Berlin, Munich, Amsterdam, Barcelona and Lisbon

12.3.1.1 Form horizon scanning to driving forces

CELTH and ETFI facilitated a workshop at Schiphol Airport for representatives from the participating cities of Lisbon, Barcelona, Berlin, Copenhagen and Amsterdam, plus a representative of ETOA. In this workshop the tourism environment was mapped and analysed. First, an introduction was given to scenario planning. Collaboratively, the participants formulated the following scenario question to guide the workshop: '*What are solutions that will relief visitor pressure to allow for a sustainable development of urban tourism in (western) European cities until 2025?*' Next, the following steps were taken subsequently:

(1) Horizon scanning. The participants were asked to identify demographic, economic, social, technological, ecological and political factors that may have a strong impact on the visitor pressure during the period until 2025 and to write them onto separate hexagons.
(2) Clustering. The participants were asked to jointly cluster the factors into internally consistent clusters.
(3) Identifying the driving forces of change. The participants were asked to identify the uncontrollable force that is believed to drive the change of each cluster and put these drivers on a separate hexagon.
(4) Reviewing the driving forces on perceived level of impact. The participants were asked to position the hexagons with the driving forces of change vertically with the strongest perceived impact at the top and the weakest perceived impact at the bottom.
(5) Reviewing the driving forces on perceived level of and unpredictability/uncertainty. The participants were asked to leave the hexagons at their vertical position and sort them horizontally, with the most predictable/certain to the left and the most unpredictable/uncertain to the right.
(6) Choosing the key uncertainties. The most uncertain driving forces of change with the strongest impact were chosen as the key uncertainties, that is, the horizontal and vertical axis of the scenario cross.
(7) Identifying plausible extremes. The participants were asked to define the limits of the plausible of both key uncertainties, or in other words, the extremes to which the key uncertainties could develop in a plausible way until 2025.

The results are shown in Figure 12.1. The analysis resulted in nine driving forces of change, of which change towards cultural understanding and 'policymaking' were identified as the two critical uncertainties, that is, those driving forces that are both the most uncertain/unpredictable and the most important. Concerning the extremes, according to the participants cultural understanding could eventually develop into the direction of full integration (cultural exchange, tourism as a means to cultural understanding, no cultural conflicts) or a situation where cultures are separated or even disintegrated. Policymaking could ultimately develop either into a strict, centrally regulated top-down process, or into a situation where policymaking is decentralised and bottom up (resident influence is stronger, more participation, more acceptance and more power to local initiatives).

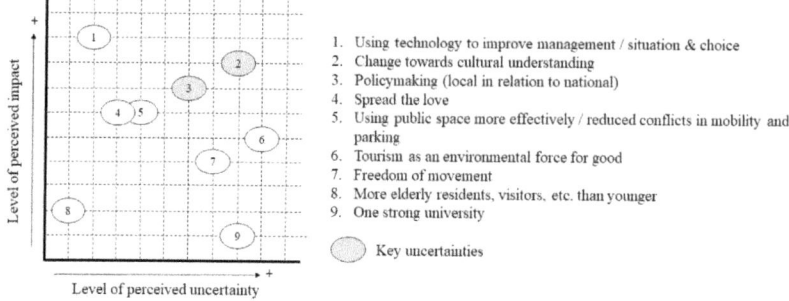

Figure 12.1 Driving forces of change and critical uncertainties for city tourism in 2025 (round 1)

When the different directions to which these key uncertainties could develop are combined, four different scenarios emerge that describe a future urban context that Destination Management Organisations (DMOs) may be faced with for the sustainable development of urban tourism in (western) European cities. The scenarios are labelled: the central city, the networked city, the atomic city and the dispersed city (see Figure 12.2).

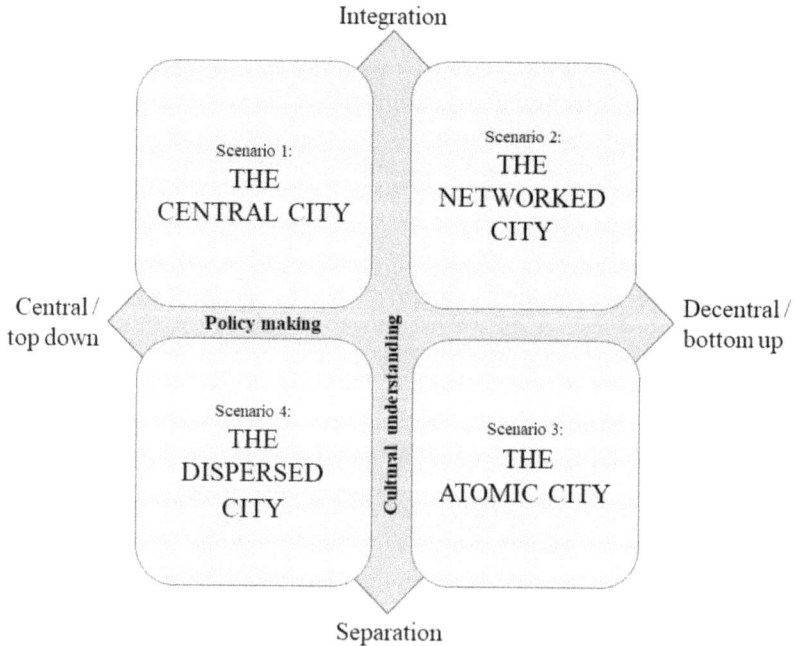

Figure 12.2 Four scenarios for the development of city tourism (round 1)

12.3.1.2 Scenarios

Below the characteristics and implications of each scenario are listed concisely.

Scenario 1: The central city

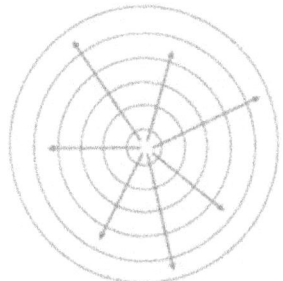

The central city is a city in which policymaking and planning is regulated centrally and top down and policy is driven from a single vision. The centralistic authorities provide the city with strict rules and regulations and with centrally managed and organised transport, which also applies to tourism and transport. Because the civil society accepts and supports the management and organisation structure of the city, there are no conflicts between different parties.

This scenario does have implications for tourism and transport. Tourism strategies benefit the entire city from a clear overarching vision on tourism in the city. This is done by either direct involvement in planning and development (government) or by shaping coherent conditions for local initiatives (subsidies, educational courses, legal regulations, zoning plans, etc.) and coordinating and guiding these initiatives (governance). Tourism to the city is primarily regarded as an economic pillar of the economy. It is used to generate income and jobs. To get as much out of tourism as possible, there is a strong commercial drive and market orientation. Everything is done to please the visitor. The product is mass touristic, well-organised, highly commoditised, coherent and 'clean'. Authenticity is staged and hospitality is standardised and impersonal. To handle the tourism flows, transferiums are located at the fringe of the city.

Scenario 2: The networked city

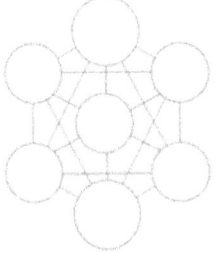

The networked city is an urban system in which multiple cultural nodes are strongly interconnected. In each node policymaking and planning is regulated locally and bottom-up. There is a strong influence of local residents and local business on policy and planning in their own environment. District governments play a coordinating role between the different actors within the district, and upwards between the districts and both within and across the nodes actors try to cooperate and to share where possible. This creates both synergy at the level of the districts and across the urban system as a whole and a cooperative, harmonious and positive atmosphere, although it is difficult to implement radical solutions.

This scenario does have implications for tourism and transport. Tourism strategies are initiated by local parties at district level, debated, discussed and adjusted in order to create synergy and benefit both the district and the city as a whole. This results in an authentic and dynamic city with a huge diversity of local flavour to be discovered and experienced. It allows the city to approach tourism not as an economic means but as a means to increase cultural understanding, and to give meaning to both the residents' and the visitors' quality of life. Local actors try their best to address the visitors personally. They understand the visitors' needs and offer them genuine hospitality. Transport is co-operatively organised by different private companies. The bottom-up and co-operative approach results in a tourism 'product' that is accessible, flexible, adaptive and resilient; in which creativity flourishes, where there is room for experiments and where innovation is ongoing.

Scenario 3: The atomic city

The atomic city represents an urban system with multiple cultural nodes that do exist side by side without any interdependency. In each node policymaking and planning is regulated locally and bottom-up. There is a strong influence of local residents and local business on policy and planning in their own environment. Although districts governments play a coordinating role between the different actors within the district, there is a fierce competition between the nodes and there is no coordination across the cultural nodes in the city.

This scenario does have implications for tourism and transport. Tourism strategies are initiated by local parties at district level, debated, discussed and adjusted in order to create synergy and benefit the own district but there is no coordination that benefits the whole city. This results in a fragmented and incoherent image of the city among (aspirant) visitors and incoherent tourist infrastructure. Different transport companies battle for the tourist, and the transport connections between the districts are fragmented. A visit to the city is a 'voyage of discovery' both during the preparation phase (marketing and promotion) and during the actual visit. There are many surprises to discover, but tourists have to be inventive to find their way. Individually the districts see tourism not only as a source of income and generator of jobs in the competition with other districts in the city, they also try to address the visitors personally by offering them authentic experiences and genuine hospitality.

Scenario 4: The dispersed city

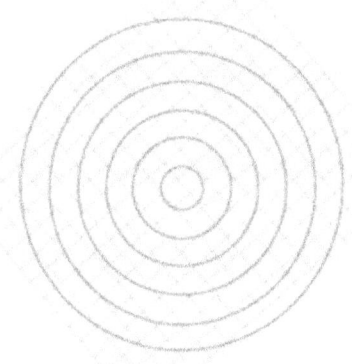

The dispersed city shows a lack of central policy, plans and rules and regulations. This causes a situation, where civil society feels uncontrolled and unbridled, and many different entities within society try to take control in order to ensure their own benefits. The society is very individualistic and competitive. There are continuous power struggles and only the strong tend to survive. The main priority is in safety and security.

In this scenario, tourism is completely overlooked and there is no strategic thinking with regard to tourism and transport whatsoever. Tourism is undeveloped. There is a lack of hospitality due to fear for 'strangers'. This also hampers creativity and innovation. Consequently, the tourist 'product' is undiscovered, unspoiled, non-commoditised, non-commercial and authentic. Such a situation is comparable to new virgin tourism destinations, where everything has to be created from scratch. The start of a (new) destination life cycle means that society slowly has to introduce initiatives to become attractive to visitors and accommodate them. This might require entrepreneurial blood from elsewhere looking for business opportunities. In order to get used to the 'strangers', such businesses could also create settings where residents and visitors can meet to develop mutual trust and understanding.

12.3.2 Round 2: Tallinn, Salzburg, Leuven, Antwerp, Ghent, Bruges and Mechelen

12.3.2.1 From horizon scanning to key uncertainties

In the previous phase of the visitor pressure study that was done in Copenhagen, Barcelona, Munich, Lisbon, Berlin and Amsterdam, the Destination Management Organisations (DMOs) collaboratively identified drivers of change and key uncertainties resulting in four scenarios. As the macro-environment is dynamic and changes continuously, driving forces of future change, key uncertainties and so the scenarios are also subject to change. During the second phase of the visitor pressure project, the steps that were conducted in round 1 were repeated for and with a new group of European cities (also at Schiphol Airport).

Just like the DMOs of the cities in the first round, the DMOs of the cities of Tallinn, Salzburg, Ghent, Antwerp, Bruges, Mechelen and Leuven have collaboratively worked on the development of a set of four scenarios. Facilitated by scenario planners at the ETFI and CELTH, they have mapped the forcefield impacting upon city tourism until 2030, have identified the two key uncertainties and have created the framework for four scenarios. The driving forces that were identified in the first round were included in the analysis. This means that the new scenario cross can be regarded as a logical evolvement of the first scenario cross, driven by changes in the environment. The scenario planners have elaborated the scenarios in four lively yet plausible storylines of 2030.

The two key uncertainties that the DMOs identified as the basis for the scenarios were:

- physical and perceived safety; and
- interaction between tourists and locals.

Five driving forces were identified, of which the development is quite predictable. However, these developments might work out differently in each of the scenarios:

- Transport related carbon pollution.
- Changing role of DMOs and policymaking.
- Overuse of local destinations and environmental impact.
- IT and data driven society.

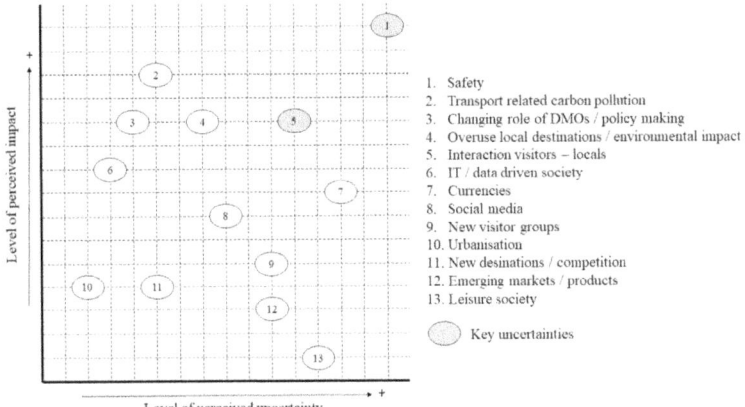

Figure 12.3 Driving forces of change and critical uncertainties for city tourism in 2025 (round 2)

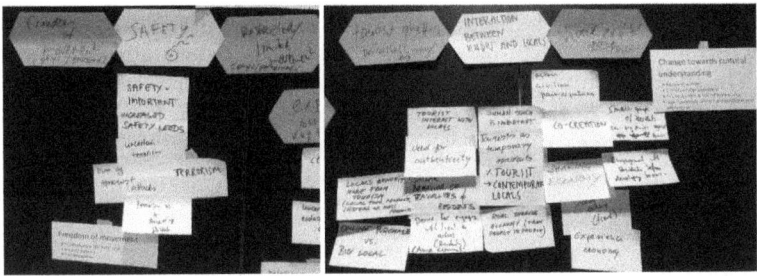

Figure 12.4 Example of two clusters of driving forces that were created during the workshop

12.3.2.2 Scenarios

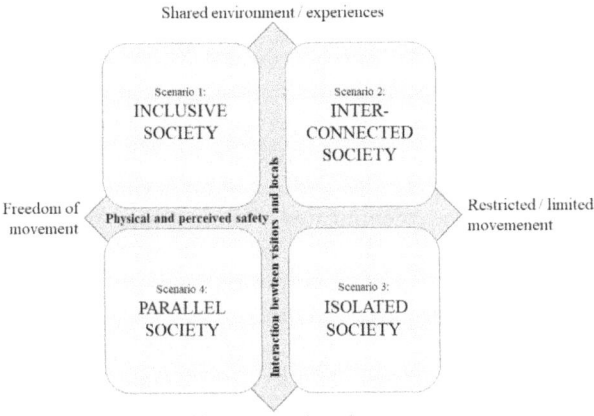

Figure 12.5 Four scenarios for the development of city tourism (round 2)

Scenario 1: Inclusive society

It is 2030. Globally, people have begun to realise that they have to work together for mankind to survive. Society has developed into an open community where differences, be it cultural, religious, political, economic, social, etc. are respected and tolerated. People value each other's differences and celebrate equality. They are happy to exchange values, beliefs, ideologies and opinions which is reflected in high levels of human interactions and high levels of personal contact, without having the fear to lose one's identity. This 'society for all' is blended and characterised by complete social integration, social inclusion and social cohesion. As it makes people proud to show to others how much they go along with others, social media are intensively used to show their altruism.

Society has developed into one global village in which people are free to travel and to visit places where they have the opportunity to engage with the locals. Fuelled by the booming experience economy, and the quest to enrich the personal quality of life, travelling between the districts of the global village has become so massive and so common, that the tourist market has become highly diversified and fragmented and includes various new visitor groups. As human relations prevail, automatisation and robotisation is limited and all kinds of actor's profit from tourism: they all contribute in their own personal way to the experience of visitors. An experience and sharing economy, pur sang. Speaking about local, regional or national identity and authenticity is something from the past. They reflect jargon from the old days.

Governments are facilitating the process in inclusiveness and tourism. The huge demand for international travel leads to massive investments in new green CO_2-free transport technology which has become cheap because of the big demand.

Scenario 2: Interconnected society

It is 2030. The tension between cultural realms of the world during the past decades, and the feelings of unsafety that are associated with it, caused the Europeans to have primarily an inward look and they are proud to experience the cultural identity they share. The inward orientation is enforced by rules and regulations that make travel across the borders of Europe complicated. These outer borders have been closed to protect Europe's identity. Yet at the national borders within Europe, passport controls were reintroduced.

The international unsafety of the past decade, and the current difficulty to travel internationally, encourage many people to stay home or to travel within their own familiar environment of the region or country, which allows locals and tourism to blend, to enjoy personal services and to share their common identity. Travel between European countries is mediocre and outside Europe extremely limited. The intra-European travel is mainly by means of train. Travelling by trains has become safer than flying, and the European high speed train network is nearly complete.

Nevertheless, governments try to restore faith in travel and tourism because they perceive tourism as a means for bridging and bonding between different nations, societies and cultures, and between locals and tourists. They view tourism as a means to create mutual understanding of each other's unique characteristics, and to learn from each other. Thus, people are encouraged to travel and to learn about other cultures beyond their own familiar environment, beyond Europe, and across the globe, to exchange ideas, beliefs, ideologies and to connect with cultures and society that are significantly different to one another.

Scenario 3: Isolated society

It is 2030. The global tension of the past decades has led to xenophobia, an irrational and/or obsessive fear of strangers, foreigners or foreign objects/cases. Because of the distrust between cultures people do not feel the need to interact and to connect with 'strangers' and the idea of environments that are shared between locals and tourists cause anxiety. Thus, the level of contact with people external to the community is minimal or non-existent.

Governments do no longer regard tourism as an engine of the economy but as a source of diseases and potential danger. There is not only fear for terrorism, but also for losing one's identity, local habits and authenticity and degradation of the environment and overuse of resources. To protect the community from external threats that are caused by tourism, visitors are unwelcome, there is ignorance to adapt to new visitor groups and the leisure economy is primarily locally oriented. The visitors who do come are preferably isolated in 'gated tourist enclaves', that is, controlled environments, yet with a high level of service. To ensure a low level of human interaction between hosts and guests, these services are primarily robotised. If people want to travel, they have to be extremely dedicated in order to overcome all the barriers and the public opinion. Since locals and visitors do not mix, the sharing economy is on the verge of extinction.

Travel takes place with conventional modes of transport and its associated levels of CO_2 emissions. Investments in transport are limited to additional safety measures.

Scenario 4: Parallel society

It is 2030. In the society at large, the age of big international tensions has ceased. However, interactions with people from other cultural groups are not valued yet and they mostly live parallel to each other without much interaction or connection. Social, spatial and intercultural contact is minimal.

Despite this, there is some fear left on the tourist market, people want to catch up with their backlog and to discover the world again, starting with the most important tangible highlights/USPs that were left at the bucket list for so many years. However, travellers are different than 10 years ago. They have become highly individualised and independent and favour individual autonomy over social inter-connectedness; they stay separated from locals and different tourist groups stay separated from each other. Technological advancements and robotised services do facilitate this attitude. The feeling that they conquered fear and travel again, gives the tourists personal satisfaction that contributes to their personal quality of life. There is great interest to share these experiences with friends, fans and followers via social media in their own parallel society.

Governments have also overcome their scruples and try to facilitate and even stimulate tourism again. The renewed interest in international tourism offers new market opportunities for tourism businesses after so many years, and they invest in new products and services and in new, affordable, accessible and fast transport technology, just to benefit from the renewed travel interest as much as possible. The investments lead to a vicious circle of considerable tourism growth, endangering the identity and authenticity of destinations and thus its core attraction value. Market driven actors across the tourism opportunity spectrum profit from the new tourism wave. The consumptive attitude of tourists and the willingness of tourism businesses to fulfil their needs, puts the core product identity and authenticity under pressure.

12.3.2.3 Scenarios and leisure lifestyles

Some initial ideas of how the scenarios relate to the tourism market may be gained by linking the scenarios to the BSR-Lifestyle model which has been developed by the Dutch hospitality industry and Smart Agent. The model uses psychographic values to classify people into different lifestyles. In the context of this project the terms were translated into English which resulted in the adapted model in Figure 12.6.

The model consists of two dimensions with which recreation behaviour can be explained: a psychological and a sociological dimension. The sociological dimension (horizontal axis) describes the extent to which one is focused on themselves (ego) or the social environment (group). The psychological dimension identifies whether a person is more extrovert or introvert towards the society. When these two dimensions are combined it results in four experience worlds. The red experience world is labelled with vitality, the yellow world with harmony, the green world with certainty and the blue world with control. Generally, people in the red and yellow worlds are looking for active forms of recreation, in the blue and green worlds

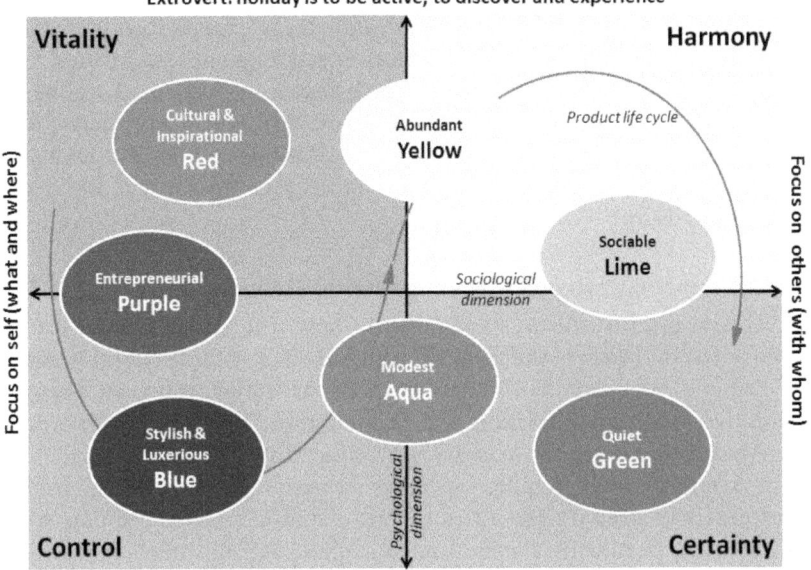

Figure 12.6 BSR lifestyle model
Source: developed by the Dutch hospitality industry and Smart Agent.

for rest and relaxation, in the red and blue worlds for cultural and sportive activities, and in the yellow and green worlds for nice VFR company.

Within the four experience worlds, seven lifestyles are positioned, each with their own profile, motives, communication channels, activity pattern, etc. At the same time, research by the Smart Agent Company shows that businesses in the early stage of their life cycle are generally oriented towards the red lifestyle, and throughout their life cycle they move from the red, via the blue and the yellow to the green lifestyles. This implies that when a business wants to rejuvenate its product this goes goad hand in hand with a repositioning on the consumer market back to the red lifestyles.

Combining the BSR lifestyle model with the scenarios suggests that certain future cities can be expected to be more attractive to people with particular lifestyles, compared to others. The *central city* can be expected to be especially attractive to blue, yellow, lime and green leisure lifestyles. This suggests that solutions within this city type are most likely to focus on cooperative solutions, which may require central leadership. The *networked city* is expected to be mainly attractive to blue, aqua and probably the yellow leisure lifestyle groups. Solutions here again can focus on cooperative solutions, but there is more room for individual interpretations and lifestyles as well. The *atomic city,* as well as the *dispersed city,* is expected to be mainly attractive to the red and purple leisure lifestyle groups. Solutions here could focus on novel experiences, which focus on individual wellbeing and specific individual aspects of the to-be-visited areas. It may prove difficult to create unity and coherence within these cities and, as such, solutions can be expected to remain more focused on the what and where dimension.

12.3.2.4 Towards strategies that anticipate the scenarios

The scenarios allow us to think how the cities could evolve when viewed through the lens of tourism, if key uncertainties move in extreme yet plausible directions. If the cities want to prepare themselves for such situations and become more future proof, it would be wise to anticipate the scenarios by choosing appropriate strategies. Each scenario will require specific strategy, yet the most robust strategies will prepare the cities for all scenarios. There are two ways in which the scenarios can be used for choosing appropriate strategic directions for the future. Either by 'wind-tunnelling' existing strategies, or by using the scenarios as a source of inspiration to develop new strategies.

12.3.2.5 Wind-tunnelling existing strategies to manage overtourism

Based on desk research within the first round of the visitor pressure study, a range of strategies was listed, with multiple measures specified per strategy. These strategies were 'wind-tunnel tested' against each of the four scenarios. The result of this wind-tunnelling exercise is shown in Table 12.1. The most robust strategies are those that take all four scenarios into account. These most robust strategies are as follows:

- Time-based rerouting (scenarios 1, 2 and 3).
- Regulation (scenarios 1, 2 and 3).
- Visitor segmentation (scenarios 1, 2 and 3).
- Communicating with and involving visitors (scenarios 1, 2 and 3).
- Communicating with and involving local stakeholders (scenarios 2, 3 and 4).

12.3.2.6 Developing new strategies that prepare for an uncertain future

Each of the scenarios has specific implications for city tourism. After the workshop with the DMOs of Tallinn, Salzburg, Ghent, Antwerp, Bruges, Mechelen and Leuven was finished, the representatives were invited to participate in a Delphi-like procedure to identify the most important direct and indirect implications of each scenario and to

Table 12.1 Robustness of the strategies for the four scenarios

		The central city	The networked city	The atomic city	The dispersed city
A	**Spreading visitors around the city and beyond** • Spread visitors to new destinations outside of the city. • Spread visitors to new destinations within the city.	+	+	o	−
B	**Time-based rerouting** • Stimulate that visitors spend more time inside tourism attractions. • Distribute visitors better during the day. • Distribute visitors better over the year.	+	+	+	−
C	**Regulation** • Prevent visitors from going to certain areas by means of transport regulations or activities. • Demotivate visitors from going to certain areas by means of higher tariffs or tourist taxes. • Create stricter rules and controls regarding the opening hours of gastronomy. • Forbid the offering of Airbnb in certain parts of the city.	+	+	+	−
D	**Creating itineraries** • Create itineraries to concentrate tourists along specific routes.	+	+	o	−
E	**Visitor segmentation** • Attract only visitors from other target groups/with other lifestyles.	+	+	+	−
F	**Make residents benefit from the visitor economy** • make residents benefit financially from visitors.	−	+	+	−
G	**Create city experiences that benefit both visitors and local residents** • create city experiences where residents and visitors can meet and integrate.	o	+	+	o
H	**Communicating with and involving visitors** • Communicate better with visitors on how to behave in the city.	+	+	+	−
I	**Communicating with and involving local stakeholders** • Communicate with and involve local residents and local businesses in tourism planning.	−	+	+	+
J	**Improve city infrastructure and facilities** • improve the infrastructure and facilities (e.g. build more roads, parking) in the city.	+	+	+/o	−

Note: The text at the bullet points refers to the way the strategies were operationalised into the survey among residents.

Table 12.2 Opportunities and threats of the four scenarios

	Major opportunities	Major threats
Inclusive society (shared environment and freedom of movement)	• Co-operation with local population. • Big data: crowd control and optimal use of spare capacity in attractions. • DMO manage instead of market.	• No diversification – more budget needed to be successful, it can win shares only by more intense advertising. • Loss of authenticity.
Interconnected society (shared environment and restricted movement)	• Start creating alliances. • Accommodation formal: Many small locally owned hotels for domestic leisure tourism. • Every town co-operates with neighbouring green areas to lengthen stay. • Job growth due to personalised services.	• Domestic: no social but physical pressure. • Shortage of international knowledge, intelligence and workforce. • International tourism: decline.
Isolated society (parallel environment and restricted movement)	• Needs to find ways to have locals explore their own city. • DMO will focus on city marketing to locals. • Raising international awareness via the internet: for example, 'dreaming of Gent' or 'dreaming of Antwerp'. • New types of businesses will rise to accommodate new needs.	• No need to internationally promote local culture – only homogeneous entertainment is requested. • Protocol takes over hosting international visitors.
Parallel society (parallel environment and freedom of movement)	• Development of specific products for every market of origin. • Promoting hidden gems for domestic market and major attractions for foreign markets. • Foster service providers to become more experimental. • Higher diversity of means of transportation.	• Dealing with complaints by locals. • No informal accommodation. • Increase of regulations to lead specification development in a certain direction.

recognise the ones perceived as a major opportunity or major threat in 2025. The opportunities and threats that were selected by more than one of the participating cities are listed in Table 12.2.[3]

For (western) European cities to become more resilient and future proof concerning visitor pressure (or overtourism) and to allow for a sustainable development of urban tourism, the challenge is to develop strategies that help to achieve opportunities and to prevent the threats. The more scenarios a strategy will cover and prepare for, the more robust it will be. In fact, such strategies would ideally prepare for all scenarios, that is, for 'any' future. Such strategies are referred to as 'robust strategies'. If strategies would prepare for only one scenario they are referred to as 'betting strategies', for two or three scenarios then they are referred to as 'semi-robust strategies'. However, the final strategy mix will also depend on how suitable, acceptable, feasible, scalable and risky they are (see Table 12.3).

In order to become adaptive and more future proof and resilient to change, strategies and tactics need to be regularly adapted to changing circumstances, that is, regularly be reviewed and reconsidered on the

Table 12.3 Evaluation criteria of potential strategies

Robustness	To how many scenarios does the strategy prepare? All: robust strategy, 2–3: semi-robust strategy, 1: betting strategy.
Suitability	To what extent does the strategy address the most important issues related to the strategic position of the organisation.
Acceptability	To what extent does the strategy align with the aims of the organisation, and to what extent is it acceptable to the stakeholders – both financial and non-financial?
Feasibility	How feasible is the strategy? To what extent could the strategic option be put in practice, in terms of resources (manpower, competences, finances, etc.), aptitude and abilities?
Scalability	To what extent is it possible to up or downscale the strategy quickly with the current personal and financial resources if market circumstances require so (market demand, competition)?
Sustainability	To what extent does the strategy align with the Sustainable Development Goals[4]?
Risk	To what extent does the strategy address future uncertainties that can affect the organisation or its strategic position negatively, or in other words: how likely is it that something goes wrong?

basis of new scenarios. There are a number of conditions for adaptivity that apply, that were discussed in Chapter 3.[5]

12.4 Concluding Remarks

In 2017, the visitor pressure study performed by CELTH and ETFI raised the interest of the World Tourism Organisation UNWTO and it was agreed that UNWTO would edit and publish the report of the first round, although it would exclude the scenario part (UNWTO, 2018). UNWTO also commissioned CELTH and ETFI to write a second volume of the report in which various city destinations around the globe reviewed their approach to manage visitor pressure and the extent to which they make use of the strategies listed in the first report (and in Table 12.1). The visitor pressure study was also disseminated into one of the first academic articles on overtourism (Koens *et al.*, 2018b) and various other publications, keynote addresses and conference presentations. Remarkably, the interest for the scenario part has been limited. Probably because the urgency of the problem of overtourism, the coping strategies gained more interest than anticipating overtourism and adaptive capacity of cities in the future.

The project showed that one of the most difficult things in managing visitor pressure is to get all relevant stakeholders on board. Particularly in cities where there are a wide variety of different perspectives, it is difficult to engage these actors in a meaningful discussion. To assist cities in this process and to create a more structured discussion, an online tool was being developed. The tool is developed on the basis of results in the first visitor pressure study and uses the management strategies

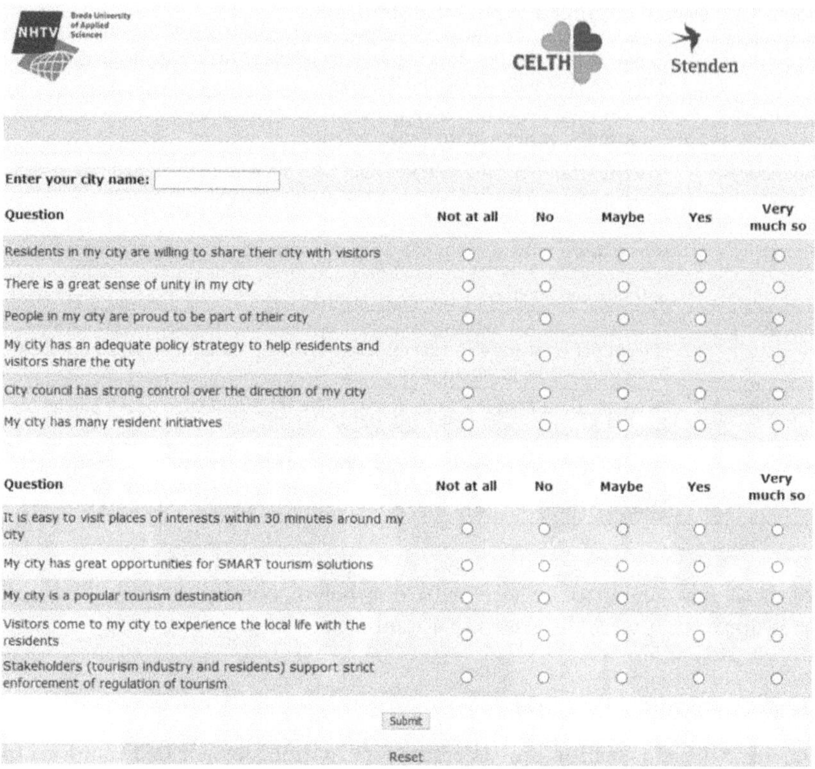

Enter your city name:

Question	Not at all	No	Maybe	Yes	Very much so
Residents in my city are willing to share their city with visitors	○	○	○	○	○
There is a great sense of unity in my city	○	○	○	○	○
People in my city are proud to be part of their city	○	○	○	○	○
My city has an adequate policy strategy to help residents and visitors share the city	○	○	○	○	○
City council has strong control over the direction of my city	○	○	○	○	○
My city has many resident initiatives	○	○	○	○	○

Question	Not at all	No	Maybe	Yes	Very much so
It is easy to visit places of interests within 30 minutes around my city	○	○	○	○	○
My city has great opportunities for SMART tourism solutions	○	○	○	○	○
My city is a popular tourism destination	○	○	○	○	○
Visitors come to my city to experience the local life with the residents	○	○	○	○	○
Stakeholders (tourism industry and residents) support strict enforcement of regulation of tourism	○	○	○	○	○

Submit

Reset

Figure 12.7 Opening questions in an online tool to structure the debate on overtourism using scenario planning

and the scenarios in particular. The tool starts with asking the different stakeholders a question regarding the city itself, as well as the role of tourism in the city (Figure 12.7), (Koens *et al.*, 2018b; UNWTO, 2018).

Based on the answers to these questions, the tool provides an indication of the perception of a person on the type of city, as well as the expected characteristics of tourism in this type of city (see Figure 12.8). This is then followed by a number of recommendations with regards to what may be the most useful management strategies in this type of city.

It is acknowledged that a tool like this is too simplistic to give a completely realistic perspective of the issues in a city. On the other hand, it does help to create a joint agenda that can stimulate different stakeholders to discuss the topic of overtourism in a coherent and structured way. As such, the results from this tool could be viewed as the starting point of a discussion for actors with, for example, a different perception on the perception of the tourism system or on how the city develops. By emphasising these differences, it may become easier for different stakeholders to come together and jointly come up with solutions.

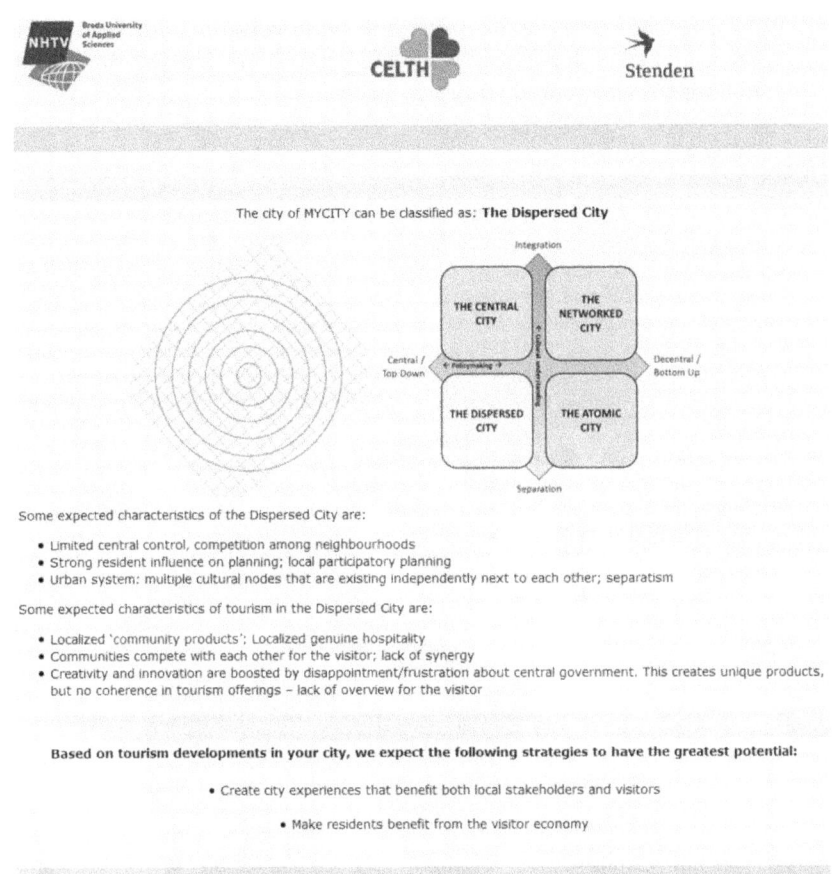

Figure 12.8 Results from an online tool to structure the debate on overtourism using scenario planning

12.5 Discussion Questions

(1) If you consider your business, organisation or destination, how is the balance between a focus on solving issues of the present and anticipating issues that may arise in the future? Could you explain why?
(2) Have you experienced any issues that could have been prevented by taking a long term view?
(3) In Figures 12.2 and 12.6 two sets of scenarios were presented. Shortlist 3 major implications for your business, organisation or destination for each of the scenarios if they would come true.
(4) Which of the scenarios would you prefer in both figures? What would be needed to direct developments in that direction?

(5) Which of the scenarios do you think will evolve in both figures? If there is a mismatch between what you prefer and what you think that will happen, what could you do to bridge this gap?

Notes

(1) CELTH is a knowledge centre that was initiated by three Universities of Applied Sciences in the Netherlands (Breda University of Applied Sciences, NHL Stenden University and HZ University of Applied Sciences), the Ministry of Education, Culture and Science and the Ministry of Economic Affairs & Climate. Other institutes and universities outside and inside the hospitality sector are also cooperated with on a project basis. CELTH's goal is to encourage public-private partnerships between universities of applied sciences and industry partners. See www.celth.nl.
(2) This network was called European Cities Marketing (ECM), recently renamed into CityDNA.
(3) Specific tables for each of the cities are also available.
(4) Accessible at https://sdgs.un.org/goals.
(5) Hartman (2017). Resilient tourism destinations? Governance implications of bringing resilience theories to tourism practice.

13 The European Tourism Futures Institute on the Edge of Time

Learning Points

- The European Tourism Futures Institute (ETFI) is the only institute in the world focusing on the future of tourism, and this chapter is a critical reflection of successes, challenges and the future.
- The interest in understanding Tourism Futures is increasing, as the industry is facing more and more pressures to adapt, such as from pandemics (COVID-19), climate change, biodiversity loss and declining social licence to operate, therefore, this book is a guide for how to manage tourism futures and scenario planning.
- The focus of ETFI's research program on scenario planning will shift more towards understanding how scenario planning can be embodied effectively by stakeholders and helping these stakeholders to do so.
- This book is only the beginning, the future has no ending.

When the idea to establish a Tourism Futures Institute was established in the late 1990s, futures thinking in tourism was rather new. Since the operations started in 2010, the European Tourism Futures Institute has increasingly contributed to developing, applying and disseminating knowledge and expertise on scenario planning and strategic foresight. In this chapter, the evolution of the ETFI is described, aligned with the stages in the tourism life cycle (Butler, 1980).

13.1 Exploration Phase: The Initiative for the European Tourism Futures Institute (2007–2011)

As explained in the foreword by Falco de Klerk Wolters, the initiative to establish the European Tourism Futures Institute (ETFI) at NHL Stenden University of Applied Sciences was rooted in a series of disruptions that affected global tourism in the first decade of the 21st century. The first ideas for the institute sprang from the mind of the management of the Academy of Leisure & Tourism, and plans were written in consultation with stakeholders in the tourism industry. The plans for a European top institute were supported with a 3 million subsidy by SNN Samenwerkingverband Noord-Nederland (Northern Netherlands Alliance) and various stakeholders, such as the province of Fryslan, province of Drenthe, City of Leeuwarden, City of Emmen and industry associations HISWA and Recron. The financial support allowed for the institute to start its operations in the autumn of 2010.

A brand identity was developed, featuring the stars of the European Union in the background to express the European focus of the newly founded institute. The globally renowned 'Internationale Tourismusbörse Berlin', commonly known as the ITB, was the stage where ETFI seized the opportunity in 2011 to present itself for the first time internationally to the international tourism industry. At that time, the field of scenario planning was rather unknown to the tourism industry and was explored by the ETFI team by means of courses provided by Future Consult (Jan Nekkers and Wybren Meijer) and the University of Houston (Andy Hines and Peter Bishop).

13.2 Involvement Phase (2011–2014)

To get the first practical experience, workshops were organised with the staff team of the Academy of Leisure & Tourism at NHL Stenden University of Applied Sciences. Also, the first scenario projects were implemented in the North of the Netherlands with regional partners, such as the city of Leeuwarden and the management of the UNESCO Geopark Hondsrug. With the aim to further develop academic knowledge in the field of tourism futures, the ETFI established the *Journal of Tourism Futures* in 2014 in close collaboration with Emerald publishers. Knowledge was shared by means of general master classes on scenario planning and specific master classes on the future outlook of branches within the leisure and tourism industry. Certificates of participation were provided to increase brand awareness and establish the ETFI's reputation. Newsletters were provided to stay in touch with the participants, increase brand awareness and continue working on the ETFI's reputation.

13.3 Development Phase (2015–2018)

While the subsidy by the SNN and partners ended in 2014, the portfolio of projects was gradually increasing in size and scope and the geographical reach of the institute increased from the Northern part of the Netherlands to the rest of the country. Increasingly, the ETFI team was invited to share their knowledge about scenario planning and their views on the future of tourism by means of performances on national and international stages (see Table 13.1). Via its keynotes, presentations,

Table 13.1 International keynotes and performances about scenario planning and scenarios in tourism

2012	World Travel Monitor Forum, Pisa, Italy
2012	Symposium Legacies of the Notting Hill Carnival, London, United Kingdom
2013	Association of Scottish Visitor Attractions ASVA, St. Andrews, Scotland
2013	ITB eTravel lab, Berlin, Germany
2014	Future Forum, Udine, Italy (*)
2014	Conference Festival Futures, Bournemouth University, Bournemouth, United Kingdom
2014	Outdoor Tourism Conference, Llanberis, Gwynedd, North Wales
2015	Scottish Camping and Caravan Forum SCCF, Edinburgh, Scotland
2015	Inaugural Attractions Management Conference, Cape Town, South Africa
2015	Chengdu University of Technology CDUT, Chengdu, China (*)
2015	International Competence Network in Tourism (ICNT) conference Sheffield, United Kingdom (*)
2016	Baltic Sea Tourism Forum, Pärnu, Estonia
2016	Urban Travel Conference, Wonderful Copenhagen, Copenhagen, Denmark
2016	International Competence Network in Tourism (ICNT) conference, Heide, Germany (*)
2017	International Conference European Cities Marketing, Dubrovnik, Croatia
2017	99th Session of the OECD Tourism Committee, OECD, Paris
2017	Chengdu University of Technology CDUT, Chengdu, China (*)
2017	ECM Annual Conference, Dubrovnik, Croatia
2018	Tourism Conference Sealand, Copenhagen, Denmark
2018	International Scientific Forum (ISF), Prague (*)
2019	International Competence Network in Tourism (ICNT) Conference, Leeuwarden (*)
2019	European Association of Tour Operators (ETOA) Conference, Luzern, Switzerland
2019	64th meeting of the UNWTO Commission for Europe (CEU), Zagreb, Croatia
2020	5th Market Intelligence Group (MIG) and Marketing Group (MG) Annual Meeting, Sliema, Malta
2021	European Cities Marketing (ECM) Autumn Conference, September 2021
2022	Sustainability Day, Expo Europe, IAAPA, Chessington, London, United Kingdom
2022	International Geographical Union (IGU) Conference, Paris, France (*)
2022	International Academic Conference on Tourism (INTACT), Yogyakarta, Indonesia (*)
2022	International Scientific Forum (ISF), Yogyakarta, Indonesia (*)
2022	International Competence Network in Tourism (ICNT) Conference, Honefoss/Oslo, Norway (*)
2022	Panel discussion at launch of OECD's report *Tourism Trends & Policies 2022*
2023	Scientific Committee on Antarctic Research (SCAR) Conference, Lisbon, Portugal (*)
2023	Futures TV, TV Channel World Federation of Futures Studies
2023	International Competence Network in Tourism (ICNT) Conference, Helsinki, Finland (*)

Notes: The table excludes presentations and performances in the Netherlands. (*) Academic conference.

projects and scenario workshops, the ETFI has been able to reach many representatives in the tourism industry and was able to increase its brand awareness and built a strong reputation.

Since the end of the start-up phase in 2014, projects that were only slightly related to scenario planning were implemented as well, such as marketing studies and visitor surveys, to cater to the practical needs of stakeholders in tourism (often becoming future partners in futures studies!), as well as to ensure financial stability of the institute as a means to carry on its futures studies.

The portfolio of scenario related studies got a boost when the ETFI initiated a pilot study on the attitude of residents towards tourism in European cities. With the support of the Centre of Expertise in Leisure, Tourism and Hospitality (CELTH), this initiative could be upscaled and the ETFI got involved with the UN World Tourism Organisation (UNWTO). The initiative resulted in a number of academic publications (e.g. Koens *et al.*, 2018b; Postma & Schmuecker, 2017; a special issue of the *Journal of Tourism Futures*) and technical reports on what was called 'overtourism' by then. The contacts with the UNWTO raised the interest of other international players, such as the World Travel & Tourism Council (WTTC), the European Association of Tour Operators (ETOA), European Cities Marketing (ECM, now called City DNA), the European Travel Commission (ETC) and the Organisation for Economic Cooperation and Development (OECD). The ETFI's initiative to conduct a post-COVID-19 scenario study for the global visitor economy (Postma *et al.*, 2020a) gave another boost to perform and share expertise in scenario planning internationally (Netherlands, France, the US, Brazil, New Zealand, Australia, tourism businesses in developing countries) although mostly online, and led to follow up studies, for example CityDNA (see Table 13.2). Over the years the portfolio of scenario studies has increased, the ETFI spread its wings more internationally, the share of multiple year projects has increased.

The development phase was also reflected in a change of management, a change in marketing and communication staff and a change in research staff. Nevertheless, throughout the years the research group on scenario planning ensured a consistent growth in the development of knowledge of and experience with scenario thinking and scenario planning, and gradually the competencies of scenario planning have been picked up by a larger share of the ETFI team and is increasingly incorporated into the DNA of the institute, as well as the Academy of Leisure & Tourism at NHL Stenden University of Applied Sciences. At the same time, scenario planning has been put into the wider context of strategic foresight, to emphasise the ETFI's ambitions to not only see the making of scenarios as a one-off effort but to incorporate scenario thinking into the strategy and planning of business, organisations and destinations.

Table 13.2 Scenario planning projects conducted by the ETFI

Year	Client	Topic
2012	Municipality of Leeuwarden	Future of the city centre of Leeuwarden 2020
2012	Chamber of Commerce, North Netherlands	Market opportunities for Groningen Airport Eelde (associated with the proclamation of the Wadden Sea as UNESCO heritage)
2012	Dutch province of Friesland	Youth and aquatics in 2020
2012	Program office UNESCO Geopark 'de Hondsrug'	Future scenarios for UNESCO Geopark 'de Hondsrug'
2013	King's Cultural Institute, London	Notting Hill Carnival Futures 2020
2013	Dutch province of Friesland	The future of overnight tourism in the province of Friesland for 2030
2013	Own initiative	The future of visitor attractions and entertainment parks in 2023
2013–2014	Industry platform of Reiswerk	The future of the travel professional in 2025
2014	Secretariat National Park Alde Feanen	Development plan National Park Alde Feanen 2022
2014	Business platform IOTO in the Dutch province of Overijssel	Increasing the innovation potential of tourism businesses in the province of Overijssel in 2025
2014	Dutch province of Noord-Holland	Tourism potential of the coastal landscapes in the North of Noord-Holland in 2050 according to four scenarios
2014	ETFI initiative	Sustainable Tourism 2040
2015	Association of group accommodation Ameland	The future of group accommodation on the island of Ameland in 2025
2015	ETFI initiative	Future scenarios for the Waddensea dike
2015	Various industry partners in the Netherlands	The future of the permanent guest in 2025
2015–2019	Destination marketing and management organisations of the cities involved	Visitor pressure in European cities; round 1; Berlin, Munich, Copenhagen, Amsterdam, Barcelona, Lisbon; round 2: Tallinn, Salzburg, Leuven, Antwerp, Ghent, Bruges and Mechelen
2017	Municipality of Emmen	Emmen 2030
2017	Dutch branch organisation of tour operators ANVR	Destination Safety in 2023
2017	NRIT media	Trend vision 2017–2020
2017	Safety Region Friesland (*)	Scenarios, policy options and roadmaps for 2027
2017–2018	Province of Friesland (*)	Future of the Frisian countryside in 2033
2018	Branche organisation Group Accommodation Netherlands	The future of group accommodation in 2035
2018	Provincie Overijssel	Vrijetijdslandschap Vechtdal
2018	Province of Limburg	The future of tourism and recreation in Limburg in 2030
2018–2019	NBTC Holland Marketing	Forecasting the development of inbound tourism till 2030

(Continued)

Table 13.2 (Continued)

Year	Client	Topic
2018–2019	Municipality of Westerveld	Future proofing indoor sport accommodation to 2030
2019	Various industry partners in the Netherlands	The future of accessible tourism in 2030
2019–2020	European Travel Commission and EU-Rail	The future of train travel 2030
2019	Lexnova (*)	The future working environment of medical specialists in Imaging and Radiation in 2030
2019–2021	Provinces bordering the IJsselmeer	The future of the recreation and tourism in the coastal area of the IJsselmeer 2050
2019–2021	Province of Friesland (*)	Futures Lab Fryslân 2030
2020	ETFI initiative	The future of the visitor economy post-COVID-19
2020	Tour operator NCRV	The future of NCRV camping travel in 3 goal-based scenarios (based on trilemma)
2021	European Cities Marketing / Tourist Information Centres	Implication of the post-COVID-19 scenarios created by CELTH in 2020
2021	Stichting Studentenstad Leeuwarden (*)	The future of Leeuwarden as a study city in 2028
2022	Branche organisation of aquatics and recreation entrepreneurs HISWA-RECRON	Transfer of the Post-COVID-19 scenarios made by CELTH in 2020 into strategic policy
2021	World Travel & Tourism Council (WTTC)	Towards Destination Stewardship: Achieving destination stewardship through scenarios and a governance diagnostics framework.
2022–2026	Dutch Research Council (NWO)	Adaptation pathways through knowledge co-production to anticipate Antarctica's uncertain tourism futures (acronym: ADAPT)
2023	City DNA	The implications of the metaverse for Destination Management Organisations in European cities
2023	Secretariat National Park Alde Feanen	From scenarios to vision on recreation and tourism in a new type of National Parks in the Netherlands
2023–2024	Province of Fryslân	Futureproofing Fryslân
2023–2024	Province of Zeeland, Center for Coastal Tourism	The visitor of the future to the province of Zeeland, The Netherlands; a baseline scenario analysis
2023–2025	European Commission	Business Events of The Future (BE-Futures)
2024	VVV Waterland van Friesland	Windtunneltesting of existing strategic plan of the Tourist Information Office against newly created scenarios, and adjusting the plan accordingly

Notes: domain of leisure & tourism, *with a component of leisure & tourism*, (*) Other domain than tourism & recreation.

13.4 Consolidation Phase (2018 to now)

Having arrived in the consolidation phase, the ETFI felt the need to share its expertise and experiences with the audience. This book is a reflection of what the institute has achieved during the past 10 years. Chapter 1 underlined the need for the establishment of a futures institute in the domain of tourism and explained how NHL Stenden University seized the opportunity to establish such an institute, with a focus on scenario planning. Chapter 2 discussed the theoretical propositions of scenario planning and provided an introduction to the usefulness and the use of scenario planning in various alternative ways. One field of application of scenario planning is to help tourism organisations, tourism businesses and tourism destinations to increase their agility and resilience and so to become more future proof. Resilience is a contemporary concept that was discussed in Chapter 3. In Chapter 4 the rise and features of two 'opposing' and complementary approaches to scenario planning were introduced. The first approach deals with a single future that can be predicted, the latter with multiple futures that can only be explored. Alternative fields of application of scenario planning were explored in Chapter 5: policy and politics, crisis management, trend analyses and product development and the design of future hotels.

The second part of the book shows how the ETFI has applied scenario planning in practice. Chapter 6 started with describing the steps in the scenario planning process as the ETFI has been applying it over the years. This is demonstrated in Chapters 7 to 12, in which a varied array of projects were presented in more detail. Chapter 7 presented the post-COVID-19 scenario study that the ETFI initiated in April 2020 with the intention to provide the industry with some future perspectives. Chapter 8 showed how the ETFI applied scenario planning to a big project concerning the future of the physical environment in the Dutch province of Friesland. Central to the project were four workshops with a large number of residents. For the administration of the same province, the ETFI was asked to set up a Futures Lab, in which all policy domains would collaborate. Due to the pandemic, the first cycle needed to be done online, with posed challenges for the scenario development process (Chapter 9). In Chapter 10, the scenario study for the Destination Marketing and Management Organisation of the Netherlands (NBTC Holland Marketing) was presented. Special to this project was the combination of the development of a predictive scenario (baseline scenario, what is assumed to happen) and alternative explorative scenarios (what could happen). The first international project, presented in Chapter 11, was commissioned by the King's Cultural Institute in London, with the aim to envision possible futures for the Notting Hill Carnival. Workshops with the carnival community, experts and visitors resulted in four scenarios. The final case that was presented focused on overtourism in several European cities. Although the focus was on mapping overtourism in these cities, the management of the destinations Management Organisations also contributed to developing scenarios for the future of city tourism. This part of the project was presented in Chapter 12.

13.5 What Next?

13.5.1 Reflective thoughts by Professor Albert Postma – Professor of Scenario Planning & Strategic Foresight

Since the ETFI was established, the work has been organised into two interdependent columns: a knowledge development led by a multiple year research programme and an implementation programme in which projects for the tourism industry are implemented. In the initial years the research programme has been focused on the macro-environment, in order to contribute to a better understanding of developments and driving forces in the demographic, economic, social, technological, environmental and political environments. The implementation programme has showcased a number of scenario planning projects (see Table 13.2). Thus, since 2010, the expertise about scenario planning and strategic foresight has increased both in theory and in practice. Although this is clearly in line with the aims, vision and mission of the ETFI, the impression is left that a really structural contribution to future proofing the tourism and recreation industry has not been achieved. Most projects seem to have an incidental character and do not lead to substantial change within the tourism and recreation industry in the way the future is anticipated. So far, most of the clients have been representing the public domain. Only a few represented business and, where this is the case, it has been mainly at the level of cooperations and industry associations. A new research programme for the period 2024–2028 aims to contribute to filling this gap.

If the ETFI wants to support the tourism industry with improving strategy development and decision making with the help of scenarios; and create more impact it is important to understand how tourism business, tourism organisations and tourism destinations want and can relate to the future. Ideally, strategic planning and scenario thinking should be intertwined and, therefore, reinforce each other. But in practice this is not very common, as it requires substantial organisational and cultural effort. Culturally, the business, organisation or region must be convinced that it performs better in the long term if it is more future oriented in thinking and acting, while thinking in scenarios must be a basic attitude. Organisationally, mechanisms of scanning and monitoring must be anchored within the organisation. This implies that signals from the future are constantly picked up, meticulously examined and assessed for their possible impacts, then compared with the scenarios in place. Scenarios need to be adjusted or renewed on a regular basis. Planning must remain flexible and connected to the dynamics of the environment, so that they form a connection between rational decision making and visionary innovation (Lindgren & Bandhold, 2009; Nekkers, 2009). In this way, the organisation works on its competence of strategic foresight, which is 'the ability to take a forward view, which enables action to be taken with reference to, and within the context of, the future' (Conway, 2007: slide 8). To achieve its ambitions, the ETFI has decided to invest more in the development of knowledge about the conditions that create the proper context for successful implementation of scenario planning and strategic foresight into the industry.

To facilitate knowledge development around cultural and organisational conditions that enhance the application of scenarios for policy and strategy development and decision making, the research group (professorship) scenario planning established a new research programme for the period 2024–2028:

- The aim is to support the industry more structurally with improving its future proofness and future resilience, by developing a better understanding of the industry at large and providing more tailormade support for individual businesses, organisations and regions that fits with the specific nature of the client in question.
- Research objectives focus on:
  - creating a better understanding of the cultural and organisational conditions of tourism businesses, tourism organisations and tourism regions;
  - developing a typology of business, organisations and regions and develop a dedicated approach to strategic foresight and scenario planning for each type;
  - developing a methodology to relate an individual client to the appropriate category; and
  - developing a tool with which the implementation of scenario planning and the future proofness can be monitored against the others in the same category.
- For each of the three categories of actors (tourism businesses, tourism organisations and tourism regions) a researcher has been appointed. They will start with conducting a literature review and make a research plan.
- Next to knowledge development by means of research, the ambition is also to strengthen the curriculum of the educational tourism programmes at our university with a stronger focus on developing competences to design and implement transitions for betterment of tourism, along professionalisation of teaching staff.
- In line with these educational ambitions, a member of the ETFI team has started a Professional Doctorate (PD) study early 2024, which is a new initiative in the Netherlands taken by CELTH and the Dutch universities of applied sciences that offer educational programmes in the field of leisure, tourism and/or hospitality. This particular study

will focus on the relation between collaborative foresight, anticipatory governance and stakeholder processes to strengthen nature-tourism destinations. The PD is a new infrastructure for education and knowledge development of applied research professionals. The aim is to allow practice-oriented research talents in the sector to contribute to innovation.

13.5.2 Reflective thoughts by Dr Stefan Hartman – Head of Department at the European Tourism Futures Institute (ETFI)

Over the last 10 years, the team of the ETFI has built an enormous track record in terms of applied science projects that revolve around scenario planning and strategic foresight. Many opportunities were presented to us by various project sponsors which allowed us to test various approaches and methodologies. Each new project has its own peculiarities and oftens needs a tailored approach. We went through many phases of design, testing and evaluation. We learned and gained experience along the way. This book contains a collection of those experiences. It will not stop here. The learning progress goes on. Here is my view on what's next.

- First, broaden the focus on tourism to the domains of leisure, events and hospitality. Being part of the overarching domain of the visitor economy, these domains face similar challenges which fit well with scenario planning methodologies.
- Second, work more often with a combination of scenario planning methodologies within one project. Already we have a number of projects in which we make combinations of a baseline scenario, a set of alternative scenarios, aspirational/goal-oriented scenarios and risk-based 'what-if' scenarios (e.g. the 'ADAPT' project on mapping a range of Antarctica's tourism futures).
- Third, strengthen the link between futures thinking and mapping alternative futures, on the one hand, and, on the other hand, the policy responses, interventions and actions that are needed at the level of businesses, organisation (public and private) and destinations. More attention towards the 'impact pathway' is needed, to ensure linkages between project output, outcomes and societal impact. This includes understanding how to integrate futures thinking and scenario planning in particular into systems of tourism destination governance.
- Fourth, integrate further scenario planning and actions geared towards managing sustainability transitions. Managing sustainability transitions is a crucial skill for future professionals in tourism. It would help to frame scenario planning firmly in the context of building adaptive capacity and its possible contribute to sustainability transitions that are urgently needed in society and for tourism to retain its social licence to operate. At the same time, it means a less descriptive approach and neutral role and a more normative, or activist, use of scenario planning.

- Fifth, use scenario planning to explore and strengthen linkages to other relevant fields of research and (policy) domains. For instance, how do the futures of tourism look like in the context of zero emission transportation, circular economy, energy neutral industries, subjective wellbeing, preventive healthcare, climate change, inclusive societies, nature conservation and place making.

All these aspects are currently being explored by members of NHL Stenden's European Tourism Futures Institute. Although we have been working with scenario planning and strategic foresight for 10 plus years, there is a bright future for applications in tourism practice ahead of us. We have only just begun.

13.5.3 Reflective thoughts by Professor Ian Yeoman – Professor of Disruption, Innovation and New Phenomena

Given the presence of societal scale risks, such as pandemics, war, climate change and AI/technology, the future of tourism will operate increasingly in a volatile, uncertain, complex and ambiguous (VUCA) world. Thus, tourism researchers will embrace tourism futures and foresight methodologies as academia becomes more focused on the future. As a consequence, it is really important to develop a stronger theoretical perspective of tourism futures as theory is the foundation of knowledge in which to frame problems, undertake research and find solutions to in an ever increasing VUCA environment in which tourism operates. My thoughts include the following:

- First, tourism academics operate in a vacuum of defending positions, often taking a narrow or singular perspective of the future. What is missing is an understanding of the interdisciplinary nature of tourism as a system. How one part of tourism effects the other. Hence, using systems thinking techniques that model the interrelationships and connectivity of tourism is vital. Systems thinking is an integral part of foresight methods in which we model possible scenarios, examine the impact of concepts between each other and determine flows, that is, cause and effect.
- Second, tourism futures must operate in a realm of scepticism. It is really important to believe in the unthinkable, the impossible and unbelievable. I am not saying that all futures have to be about science fiction, but they need a degree of science fiction. A good scenario set of four scenarios must have a least one or two sceptical scenarios as the question at the centre of scenario planning is 'what if'. It is about explaining or understanding how the sceptical scenarios could come true.
- Third, building on the platform of scepticism, is ensuring that scenario sets move beyond rationality and reasoning. Scenarios must not look similar, hence, diversity is important. This means scenarios are embedded

in Dator's Alternative Futures or (Dator, 1981), Bergman's Future States (Bergman *et al.*, 2010) or other models of ontological plurality.
- Fourth, given a VUCA environment of tourism's future, I see tourism in academia further embracing scenario planning as a theoretical framework, when researching tourism. Scenario planning is based upon ontological plurality in which conceptualisation is based upon a 2 × 2 matrix with four scenarios which represent ideas, states, futures, theories or research. This approach is aligned with Pearce's Tourism Research Frameworks approach (Pearce, 2012). In particular, this approach is useful for PhD research, in which students can engage a multiple futures perspective.
- Fifth, further work is still required in developing paradigms of tourism futures within a hierarchy of ontology, epistemology and methodology. Futures paradigms unique to tourism are envisaged. For example, the future past paradigm (Yeoman & McMahon-Beattie, 2020) is a reflection that tourism's future is based on culture and history as destinations icons and geography fundamentally do not change, whereas, a disruptive paradigm is based upon a radical and rapid change future which could be brought about technology (Yeoman *et al.*, 2021).

13.6 Concluding Remarks

In conclusion, we asked ChatGTP (on 14 December 2023) 'what is the role of scenario planning in predicting the future of tourism' and the answer was:

Scenario planning is a strategic management tool that involves creating and considering multiple plausible future scenarios to help organizations prepare for uncertainty and make more informed decisions. In the context of predicting the future of tourism, scenario planning plays a crucial role in navigating the complexities and uncertainties inherent in the tourism industry. Here are some ways in which scenario planning can be valuable:

- *Anticipating Changes in the Tourism Landscape: Tourism is influenced by a wide range of factors such as economic conditions, geopolitical events, technological advancements, and environmental changes. Scenario planning helps tourism stakeholders identify potential drivers of change and understand how they might interact to shape the future.*
- *Risk Management: By developing different scenarios, tourism planners can assess the potential risks and uncertainties that may arise. This allows for the development of strategies to mitigate risks and capitalize on opportunities in various future scenarios.*
- *Strategic Decision Making: Tourism organizations can use scenario planning to inform their strategic decision-making processes. By exploring different possible futures, they can identify strategies that are robust across multiple scenarios, helping them make more resilient and adaptable plans.*

- *Informed Policy Development:* Governments and tourism authorities can use scenario planning to inform the development of policies that support a sustainable and resilient tourism industry. This may involve preparing for changes in regulations, infrastructure development, or destination management strategies.
- *Adaptation to Market Trends:* Tourism businesses can use scenario planning to anticipate and adapt to changing consumer preferences, technological advancements, and market trends. This allows them to stay ahead of the curve and remain competitive in a rapidly evolving industry.
- *Collaborative Planning:* Scenario planning often involves collaboration among various stakeholders, including government bodies, businesses, and community representatives. This collaborative approach ensures a more comprehensive understanding of potential future developments and fosters collective efforts to address challenges and opportunities.
- *Enhanced Preparedness:* The tourism industry is susceptible to unexpected events such as natural disasters, public health crises, or geopolitical conflicts. Scenario planning enables organizations to enhance their preparedness for such events by considering a range of possible scenarios and developing contingency plans.
- *Long-Term Vision:* Scenario planning encourages a long-term perspective, allowing tourism stakeholders to envision and prepare for developments that may unfold over an extended period. This helps in developing sustainable strategies that consider the long-term impact of decisions.

In summary, scenario planning in the context of tourism helps stakeholders to be proactive, adaptive, and resilient in the face of uncertainties. It enables them to explore various future possibilities, understand potential challenges and opportunities, and make informed decisions to shape the future of tourism in a dynamic environment.

Thus, the second edition will be written by artificial intelligence rather than the authors (scary thought), so is this the beginning of the end or a depiction of the future taken from the storyline of Terminator (Cameron *et al.*, 1998)?

13.7 Discussion Questions

(1) In your own opinion, what is the future for scenario planning in shaping the debate about the future of tourism?
(2) Reflecting upon this book, what are your key learning points?
(3) If we were to write the second edition of this book, what should be in the book that is not in this edition?
(4) You have read the book, so what are your scenarios for the future of tourism?

References

Ackoff, R.L. (1978) *The Art of Problem Solving*. Wiley.

Adam, B. and Groves, C. (2007) *Future Matters: Action, Knowledge, Ethics*. Brill.

ADB (2021) *Looking Forward Vol 1: Evaluating the Challenges for the Pacific Tourism After COVID-19* (Pacific Tourism Sector Assessment, Issue. A. D. Bank).

Albrecht, J.N. (2017) Challenges in national-level tourism strategy implementation – A long-term perspective on the New Zealand tourism strategy 2015. *International Journal of Tourism Research* 19 (3), 329–338. https://doi.org/10.1002/jtr.2115.

Amara, R. (1974) The futures field: Functions, forms and critical issues. *Futures* 6 (4), 289–301. https://doi.org/10.1016/0016-3287(74)90072-X.

Amara, R. (1981) The futures field: Searching for definitions and boundaries. *The Futurist* 15 (1), 25–9.

Amer, M., Daim, T.U. and Jetter, A. (2013) A review of scenario planning. *Futures* 46, 23–40. https://doi.org/10.1016/j.futures.2012.10.003.

Asselt, M., Klooster, S., Notten, P. and Smits, L. (2010) *Foresight in Action: Developing Policy-Oriented Scenarios*. Routledge.

Axelrod, R. and Cohen, M. (2000) *Harnessing Complexity: Organisational Implications of a Scientific Frontier*. Basic Books.

Baggio, R. (2008) Symptoms of complexity in a tourism system. *Tourism Analysis* 13 (1), 1–20.

Ball, P., Vince, G., Kucharski, A., Prasad, A., Rutherford, A., Climer, N. and Woodward, A. (2017) *What's Next?: Even Scientists Can't Predict the Future–or Can They?* Profile Books.

Bates, D.G. and Tucker, J. (2010) *Human Ecology: Contemporary Research and Practice*. Springer. https://doi.org/10.1007/978-1-4419-5701-6.

Becken, S. and Loehr, J. (2022) Asia–Pacific tourism futures emerging from COVID-19 recovery responses and implications for sustainability. *Journal of Tourism Futures* (ahead-of-print). https://doi.org/10.1108/JTF-05-2021-0131.

Bentley, T. (1994) Facilitation: Providing opportunities for learning. *Journal of European Industrial Training* 18 (5), 8–22.

Berger, G. (1957) Sciences humaines et prevision [Social Sciences and Forecasting]. *La Revue des Deux Mondes* 1 (3), 16–26.

Berger, P.L. and Luckmann, T. (1967) *The Social Construction of Reality: A Treatise in the Sociology of Knowledge*. Penguin.

Bergman, A., Karlsson, J.C. and Axelsson, J. (2010) Truth claims and explanatory claims – An ontological typology of futures studies. *Futures* 42 (8), 857–865. https://doi.org/10.1016/j.futures.2010.02.003.

Bhaskara, G.I. and Filimonau, V. (2021) The COVID-19 pandemic and organisational learning for disaster planning and management: A perspective of tourism businesses from a destination prone to consecutive disasters. *Journal of Hospitality and Tourism Management* 46, 364–75.

Bishop, P., Hines, A. and Collins, T. (2007) The current state of scenario development: An overview of techniques. *Foresight* 9 (1), 5–25. https://doi.org/10.1108/14636680710727516.

Blass, E. (2003) Researching the future: Method or madness? *Futures* 10, 1041–1054.

Boschma, R. (2015) Towards an evolutionary perspective on regional resilience. *Regional Studies* 49 (5), 733–751. https://doi.org/10.1080/00343404.2014.959481.

Börjeson, L., Höjer, M., Dreborg, K.-H., Ekvall, T. and Finnveden, G. (2006) Scenario types and techniques: Towards a user's guide. *Futures* 38 (7), 723–739. https://doi.org/10.1016/j.futures.2005.12.002.

Boston, J. (2017) *Safeguarding the Future Governing in an Uncertain World*. Bridget Williams Books.

Bradfield, R. (2008) Cognitive barriers in the scenario development process. *Advances in Developing Human Resources* 10 (2), 198.

Bradfield, R., Wright, G., Burt, G., Cairns, G. and Van Der Heijden, K. (2005) The origins and evolution of scenario techniques in long range business planning. *Futures* 37 (8), 795–812. https://doi.org/10.1016/j.futures.2005.01.003.

Brouder, P. and Eriksson, R.H. (2013) Tourism evolution: On the synergies of tourism studies and evolutionary economic geography. *Annals of Tourism Research* 43, 370–389.

Butler, R.W. (1980) The concept of a tourist area cycle of evolution: Implications for management of resources. *Canadian Geographer* 24, 5–12.

Callaghan, V., Miller, J., Yampolskiy, R. and Armstrong, S. (2017) *The Technological Singularity Managing the Journey*. Springer.

Cameron, J., Schwarzenegger, A., Hamilton, L. and Patrick, R. (1998) *The Terminator* (MAWA Film & Medien).

Carnival Futures, Online information, King's College London, King's Culture. See https://www.kcl.ac.uk/cultural/projects/2016/carnival-futures (accessed June 2022).

Carnival Futures: Notting Hill Carnival 2020 Video. See https://youtu.be/ef4jAXXQu8Q (accessed June 2022).

CELTH (2017) Trend visie 2017–2020. *Recreatie and Toerisme* December 2017, 22–29. https://www.retailinsiders.nl/docs/3fb4bae8-5805-40d6-9088-41ed108a6ff1.pdf.

Checkland, P. (1990) *Soft Systems Methodology in Action*. Wiley.

Chermack, T.J. (2004) Improving decision-making with scenario planning. *Futures* 36 (3), 295–309. https://doi.org/10.1016/S0016-3287(03)00156-3.

Chermack, T.J. (2005) Studying scenario planning: Theory, research suggestions and hypotheses. *Technological Forecasting and Social Change* 72 (1), 59–73. https://doi.org/10.1016/j.techfore.2003.11.003.

Chermack, T.J. (2007) Disciplined imagination: Building scenarios and building theories. *Futures* 39 (1), 1–15. https://doi.org/10.1016/j.futures.2006.03.002.

Chermack, T.J. and Coons, L.M. (2015) Scenario planning: Pierre Wack's hidden messages. *Futures* 73, 187–193. https://doi.org/10.1016/j.futures.2015.08.012.

Chermack, T.J. and van Der Merwe, L. (2003) The role of constructivist learning in scenario planning. *Futures* 35 (5), 445–460. https://doi.org/10.1016/S0016-3287(02)00091-5.

Chermack, T.J. and Walton, J.S. (2006) Scenario planning as development and change intervention. *International Journal of Agile Systems and Management* 1 (1), 46–59.

Chermack, T.J., Lynham, S.A. and Ruona, W.E. (2001) A review of scenario planning literature. *Futures Research Quarterly* 17 (2), 7–32.

Chermack, T., Coons, L., Nimon, K., Bradley, P. and Glick, M. (2015) The effects of scenario planning on participant perceptions of creative organizational climate. *Journal of Leadership and Organizational Studies* 22 (3), 355.

Clark, C., Nyaupane, G.P., Timothy, D.J. and Buzinde, C. (2022) Scenario planning as a tool to manage tourism uncertainties during the era of COVID-19: A case study of Arizona, USA. *Current Issues in Tourism* 25 (7), 1063–1073.

Conway, M. (2006) An overview of foresight methodologies. *Thinking Futures* 10, 1–10.

Conway, M. (2007) Building a strategic foresight capacity. Presentation at Australasian Association for Institutional Research (AAIR) Forum, Brisbane, QLD, November 2007.

Cooperrider, D.L. and Whitney, D. (2005) *Appreciative Inquiry. A Positive Revolution in Change*. Berrett-Koehler Publishers, Inc.

Cordova-Pozo, K. and Rouwette, E.A.J.A. (2023) Types of scenario planning and their effectiveness: A review of reviews. *Futures* 149, 103153. https://doi.org/10.1016/j.futures.2023.103153.

Cvelbar, L. and Ogorevc, M. (2020) Saving the tourism industry with staycation vouchers [version 1; peer review: 2 approved, 1 approved with reservations]. *Emerald Open Research* 2 (65). https://doi.org/10.35241/emeraldopenres.13924.1.

Dator, J. (1971) Dimensions of the future: Washington, May 1971 First General Assembly of the World Future Society 3, 311–113.

Dator, J.I. (1981) Alternative futures and the futures of law. In D. James and C. Bezold (eds) *Judging the Future*. University of Hawaii Press.

Dator, J. (2009) Alternative futures at the Manoa School. *Journal of Futures Studies* 1 (2), 1–18.

Dator, J. (2015) *Mutative Media: Communication Technologies and Power Relations in the Past, Present and Futures*. Springer International Publishing. https://doi.org/10.1007/978-3-319-07809-0.

Davoudi, S. (2012) Resilience: A bridging concept or a dead end? *Planning Theory and Practice* 13 (2), 299–307.

De Geus, A. (1988) Planning as learning. *Harvard Business Review* March-April, 70–74.

De Jouvenel, H. (2018) Futuribles: Origins, philosophy, and practices – Anticipation for action. *World Futures Review* 11 (1), 8–18. https://doi.org/10.1177/1946756718777490.

De Roo, G. (2012) Spatial planning, complexity and a world 'Out of equilibrium': Outline of a nonlinear approach to planning. In G. De Roo, J. Hillier and J. Van Wezemael (eds) *Complexity and Planning: Systems, Assemblages and Simulations* (pp. 129–165). Ashgate.

De Roo, G. and Boelens, L. (eds) (2014) *Spatial Planning in a Complex Unpredictable World of Change: Towards a Proactive Co-evolutionary Type of Planning within the Eurodelta*. InPlanning.

Derbyshire, J. (2016) The implications, challenges and benefits of a complexity-orientated. Futures Studies. *Futures* 77, 45–55. https://doi.org/10.1016/j.futures.2016.02.001.

Derbyshire, J. (2017) Potential surprise theory as a theoretical foundation for scenario planning. *Technological Forecasting and Social Change* 124, 77–87.

Dredge, D. and Jamal, T. (2015) Progress in tourism planning and policy: A post-structural perspective on knowledge production. *Tourism Management* 51, 285–297. https://doi.org/10.1016/j.tourman.2015.06.002.

Dredge, D. and Jenkins, J.M. (2007) *Tourism Planning and Policy*. John Wiley and Sons.

Dredge, D. and Jenkins, J.M. (2011) *Stories of Practice Tourism Policy and Planning*. Ashgate Publishing Company.

Dubin, R. (1978) *Theory Building*. Free Press.

Durance, P. and Godet, M. (2010) Scenario building: Uses and abuses. *Technological Forecasting and Social Change* 77 (9), 1488–1492. https://doi.org/10.1016/j.techfore.2010.06.007.

Eden, C. and Ackerman, F. (1998) *Making Strategy: The Journey of Strategic Management*. Sage Publications.

European Foresight Platform (2022) *Environmental Scanning*. See http://www.foresight-platform.eu/community/forlearn/how-to-do-foresight/methods/analysis/environmental-scanning/ (accessed June 2022).

European Foresight Platform (n.d.) *What is Foresight?* See http://foresight-platform.eu/community/forlearn/what-is-foresight/ (accessed April 2022).

European Tourism Futures Institute (ETFI) (2018) *Scenariostudie Limburg 2030: Bouwen aan een toekomstvaste sector Toerisme en Recreatie*. European Tourism Futures Institute.

Evans, S.K. (2011) Connecting adaptation and strategy: The role of evolutionary theory in scenario planning. *Futures* 43 (4), 460–468.

Fabry, N. and Zeghni, S. (2019) Resilience, tourist destinations and governance: An analytical framework. In F. Cholat, L. Gwiazdzinski, C. Tritz and J. Tuppen (eds) *Tourismes et Adaptations* (pp. 96–108), 9791091336123. Elya Editions (hal-02070497).

Factory, F. (2018a) *Digital Revoltion: Contextual Driver*. Foresight Factory.
Factory, F. (2018b) *The Experience Seeker*. Foresight Factory.
Factory, F. (2018c) *Mercurial Consumption*. Foresight Factory.
Farsari, I., Butler, R.W. and Szivas, E. (2011) Complexity in tourism policies: A cognitive mapping approach. *Annals of Tourism Research* 38 (3), 1110–1134. https://doi.org/10.1016/j.annals.2011.03.007.
Ferdinand, N. (2013) *Creative Futures Project Proposal and Delivery Plan*. King's College, London.
Ferdinand, N. and Williams, N.L. (2018) The making of the London Notting Hill Carnival festivalscape: Politics and power and the Notting Hill Carnival. *Tourism Management Perspectives* 27 (July), 33–46. https://doi.org/10.1016/j.tmp.2018.04.004.
Fergnani, A. and Chermack, T.J. (2021) The resistance to scientific theory in futures and foresight, and what to do about it. *Futures and Foresight Science* 3, 3–4. https://doi.org/10.1002/ffo2.61.
Flatters, P. and Willmott, M. (2009) Understanding the post-recession consumer. (Special Issue: Managing in the New World.) *Harvard Business Review* 87 (7–8), 106.
Gangwer, T. (2009) *Visual Impact, Visual Teaching: Using Images to Strengthen Learning* (2nd edn). Corwin Press. http://www.loc.gov/catdir/toc/ecip0824/2008031922.html.
Gavetti, G. (2012) Perspective—Toward a behavioral theory of strategy. *Organization Science* 23 (1), 267–285. https://doi.org/10.1287/orsc.1110.0644.
Gershwin, L.-a. and Crowley-Cyr, L. (2021) Forecasting hazardous jellyfish: Shifting perceptions from black swans events to white. In G.L. Marittini, N. Killi and L. Xiao (eds) *The Cnidaria: Only a Problem or also a Resource?* (pp. 123–140). Nova Science Publishers.
Gidley, J. (2017) *The Future: A Very Short Introduction*. Oxford University Press.
Godet, M. and Durance, P. (2011) *Strategic Foresight for Corporate and Regional Development*. UNESCO.
Google Arts and Culture (n.d.) A history of Notting Hill carnival. Exploring the history and politics of the development of Notting Hill carnival. See https://artsandculture.google.com/story/a-history-of-notting-hill-carnival-black-cultural-archives/pwXhDoHj8Po6Kg?hl=en (accessed September 2021).
Gordon, A.V. (2020) Matrix purpose in scenario planning: Implications of congruence with scenario project purpose. *Futures* 115, 102479.
Gram, M., O'Donohoe, S., Schänzel, H., Marchant, C. and Kastarinen, A. (2019) Fun time, finite time: Temporal and emotional dimensions of grand travel experiences. *Annals of Tourism Research* 79, 102769. https://doi.org/10.1016/j.annals.2019.102769.
Gurdjieff, G.I. (1960) *Meeting with Remarkable Men*. Penguin.
Habermas, J. (1974) *Theory and Practice* (J. Viertel, Trans.). Heinemann.
Hales, J. (1969) Futures: Confidence from chaos. *Futures* 1 (1), 2–3. https://doi.org/10.1016/S0016-3287(69)80001-7.
Hall, C.M., Lundmark, L. and Zhang, J. (eds) (2021) *Degrowth and Tourism: New Perspectives on Tourism Entrepreneurship, Destinations and Policy*. Routledge.
Hamel, G. and Pralahad, C.K. (2005, 3rd edn.) *La conquête du futur. Construire l'avenir de son entreprise plutôt que le subir*. Dunod.
Hartman, S. (2016) Towards adaptive tourism areas? A complexity perspective to examine the conditions for adaptive capacity. *Journal of Sustainable Tourism* 24 (2), 299–314. https://doi.org/10.1080/09669582.2015.1062017.
Hartman, S. (2018a) Towards adaptive tourism areas: Using fitness landscapes for managing and futureproofing tourism area development. *Journal of Tourism Futures* 4 (2), 152–162. https://doi.org/10.1108/JTF-03-2018-0009.
Hartman, S. (2018b) Resilient tourism destinations? Governance implications of bringing theories of resilience and adaptive capacity to tourism practice. In E. Innerhofer, M. Fontanari and H. Pechlaner (eds) *Destination Resilience – Challenges and*

Opportunities for Destination Management and Governance (pp. 66–75). Routledge. https://doi.org/10.4324/9780203701904.

Hartman, S. (2021) Adaptive tourism areas in times of change. *Annals of Tourism Research* 87. https://doi.org/10.1016/j.annals.2020.102987.

Hartman, S. (2023) Destination governance in times of change: A complex adaptive systems perspective to improve tourism destination development. *Journal of Tourism Futures* 9 (2), 267–278. https://doi.org/10.1108/JTF-11-2020-0213.

Hartman, S., Wielenga, B. and Heslinga, J. (2020) The future of tourism destination management: building productive coalitions of actor networks for complex destination development. *Journal of Tourism Futures* 6 (3), 213–218. https://doi.org/10.1108/JTF-11-2019-0123.

Hartman, S., Heslinga, J., Oskam, J., Revier, H., De Vries, M. and Zandberg, T. (2014) *De Waddendijk. Mogelijkheden voor een vrijetijdslandschap*. European Tourism Futures Institute.

Hay, B. (2019) The future of national tourism organisations marketing functions – there is no future? *Journal of Tourism Futures* (ahead-of-print). https://doi.org/10.1108/JTF-08-2019-0075.

Hay, B. (2021) Reflections on the future visions of UK tourism outlined in Burkart and Medlik's 1974 book: Tourism: past, present, and future. *Journal of Tourism Futures* (ahead-of-print). https://doi.org/10.1108/JTF-11-2020-0217.

Heylighen, F. (2001) The science of self-organization and adaptivity. In L.D. Kiel (ed.) *The Encyclopedia of Life Support Systems* (pp. 253–280). Eolss.

Hideg, E. (2002) Implications of two new paradigms for futures studies. *Futures* 34 (3–4), 283–294. https://doi.org/10.1016/S0016-3287(01)00044-1.

Hideg, E. (2013) Integral futures based on the paradigm approach. *Futures* 45 (1), 6–15. https://doi.org/10.1016/j.futures.2012.11.007.

Hiltunen, E. (2008) Good sources of weak signals: A global study of where futurist look for weak signals. *Journal of Futures Studies* 12 (4), 21–44.

Hines, A. and Bishop, P. (2015) *Thinking About the Future. Guidelines for Strategic Foresight* (2nd edn). Hinesight.

Hodgkinson, G.P. and Clarke, I. (2007) Conceptual note: Exploring the cognitive significance of organizational strategizing: A dual-process framework and research agenda. *Human Relations* 60 (1), 243–255. https://doi.org/10.1177/0018726707075297.

Holton, D. and Clarke, D. (2006) Scaffolding and metacognition. *International Journal of Mathematical Education in Science and Technology* 37 (2), 127–143. https://doi.org/10.1080/00207390500285818.

Huff, A.S. and Jenkins, M. (2002) *Mapping Strategic Knowledge*. Sage. https://doi.org/http://site.ebrary.com/lib/vuw/Doc?id=10080940.

Hussain, A. (2022) Human mobility, hospitality, and tourism industries: A perspective on catastrophes. *Journal of Sustainability and Resilience* 2 (2), 4.

Inayatullah, S. (1998) Causal layered analysis: Poststructuralism as method. *Futures* 30 (8), 815–829. See https://web.archive.org/web/20140714133227/http:/proutglobal.info/slideshows/training/Related%20web%20pages/002-Casual_Layered_Analysis.pdf (accessed December 2021).

Inayatullah, S. (2010) Theory and practice in transformation: The disowned futures of integral extension. *Futures* 42 (2), 103–109. https://doi.org/10.1016/j.futures.2009.09.002.

Kahn, H. and Wiener, A.J. (1967) *The Year 2000: A Framework for Speculation on the Next Thirty-Three Years*. Macmillan.

Karlsen, J. E. and Karlsen, H. (2013) Classification of tools and approaches applicable in foresight studies. In M. Giaoutzi and B. Sapio (eds) *Recent Developments in Foresight Methodologies* (pp. 27–52). Springer. https://doi.org/10.1007/978-1-4614-5215-7_3.

Karlsen, J.E., Øverland, E.F. and Karlsen, H. (2010) Sociological contributions to futures' theory building. *Foresight* 12 (3), 12. http://search.proquest.com.helicon.vuw.ac.nz/docview/357276405?OpenUrlRefId=info:xri/sid:primo&accountid=14782.

Kauffman, S.A. (1993) *Origins of Order: Self Organization and Selection in Evolution*. Oxford University Press.
Keough, S.M. and Shanahan, K.J. (2008) Scenario planning: Toward a more complete model for practice. *Advances in Developing Human Resources* 10 (2), 166–178. https://doi.org/10.1177/1523422307313311.
Kerlinger, F.N. and Lee, H.B. (2000) *Foundations of Behavioral Research* (4th edn). Harcourt College.
Klein, G., Moon, B. and Hoffman, R.R. (2006) Making sense of sensemaking 1: Alternative perspectives. *IEEE Intelligent Systems* 21 (4), 70–73.
Koens, K. and Postma, A. (2017, unpublished) Understanding and Managing Visitor Pressure in Urban Tourism. A Study to into the Nature of and Methods used to Manage Visitor Pressure in Six Major European Cities. Breda/Leeuwarden/Vlissingen: CELTH (processed into UNWTO, 2018).
Koens, K., Postma, A. and Papp, B. (2018a, April) Managing visitor pressure in European destinations. Paper presented at the *12th International Scientific Forum*. Prague, Czech Republic, 26–27 April 2018.
Koens, K., Postma, A. and Papp, B. (2018b) Is overtourism overused? Understanding the impact of tourism in a city context. *Sustainability* 10 (12), 4384. https://www.mdpi.com/2071-1050/10/12/4384/htm).
Kolb, D.A. (1984) *Experiential Learning* (pp. 149–159). Prentice Hall.
Kurzweil, R. (2005) *The Singularity is Near : When Humans Transcend Biology*. Viking.
Lave, J. and Wenger, E. (1991) *Situated Learning: Legitimate Peripheral Participation*. Cambridge University Press.
Lee, L.J. (2019) Tools: MIRO Real-time board, visual collaborations and tools, easy screen sharing and presentation. Presentation. MELSpace, Monitoring, Evaluation and Learning Repository, Agricultural Research Knowledge. See https://hdl.handle.net/20.500.11766/10605 (accessed June 2022).
Levi-Strauss, C. (1966) *The Savage Mind: (La Pensee Sauvage)*. Weidenfeld and Nicolson.
Lew, A. (2014) Scale, change and resilience in community tourism planning. *Tourism Geographies* 16 (1), 14–22. https://doi.org/10.1080/14616688.2013.864325.
Lindgren, M. and Bandhold, H. (2009) *Scenario Planning: The Link Between Future and Strategy*. Palgrave Macmillan Limited. https://doi.org/10.1057/9780230511620.
Linstone, H.A. and Turoff, M. (2011) Delphi: A brief look backward and forward. *Technological Forecasting and Social Change* 78 (9), 1712–1719. https://doi.org/10.1016/j.techfore.2010.09.011.
López-Rodríguez, M.D., Oteros-Rozas, E., Ruiz-Mallén, I., March, H., Horcea-Milcu, A.I., Heras, M., Cebrián-Piqueras, M.A., Andrade, R., Lo, V.B. and Piñeiro, C. (2023) Visualizing stakeholders' willingness for collective action in participatory scenario planning. *Ecology and Society* 28 (2), 5.
Lub, X.D., Rijnders, R., Caceres, L.N. and Bosman, J. (2016) The future of hotels: The Lifestyle Hub. A design-thinking approach for developing future hospitality concepts. *Journal of Vacation Marketing* 22 (3), 249–264. https://doi.org/10.1177/1356766715623829.
Lubowiecki-Vikuk, A. and Sousa, B. (2021) Tourism business in a vuca world: Marketing and management implications. *Journal of Environmental Management and Tourism* 12 (4), 867–876. https://doi.org/10.14505//jemt.v12.4(52).01.
Lubowiecki-Vikuk, A., Budzanowska-Drzewiecka, M., Borzyszkowski, J. and Taheri, B. (2023) Critical reflection on VUCA in tourism and hospitality marketing activities. *International Journal of Contemporary Hospitality Management* 35 (8), 2983–3005. https://doi.org/10.1108/IJCHM-04-2022-0479.
Lugosi, P., and Ndiuini, A. (2022) Migrant mobility and value creation in hospitality labour. *Annals of Tourism Research* 95, 103429.
Luhmann, N. (1976) The future cannot begin: Temporal structures in modern society. *Social Research* 43 (1), 130–152.

Luo, X. (2002) Uses and gratifications theory and e-consumer behaviors: A structural equation modeling study. *Journal of Interactive Advertising* 2 (2), 34–41.

Lynham, S.A. (2000) Theory building in the human resource development profession. *Human Resource Development Quarterly* 11 (2), 159–178. https://doi.org/10.1002/1532-1096(200022)11:2<159::AID-HRDQ5>3.0.CO;2-E.

Ma, M. and Hassink, R. (2013) An evolutionary perspective on tourism area development. *Annals of Tourism Research* 41, 89-109.

Mackay, B. and Tambeau, P. (2013) A structuration approach to scenario praxis. *Technological Forecasting and Social Change* 80 (4), 673.

Major, J. and Clarke, D. (2022) Regenerative tourism in Aotearoa New Zealand – A new paradigm for the VUCA world. *Journal of Tourism Futures* 8 (2), 194–199. https://doi.org/10.1108/JTF-09-2021-0233.

Mannermaa, M. (1991) In search of an evolutionary paradigm for futures research. *Futures* 23 (4), 349–372. https://doi.org/10.1016/0016-3287(91)90111-E.

Marchais-Roubelat, A. and Roubelat, F. (2008) Designing action-based scenarios. *Futures* 40 (1), 25–33. https://doi.org/10.1016/j.futures.2007.06.008.

Martelli, A. (2014a) *Models of Scenario Building and Planning*. Brocconi University Press/Palgrave MacMillan.

Martelli, A. (2014b) *Models of Scenario Building and Planning: Facing Uncertainty and Complexity*. Palgrave Macmillan.

Martin, B.R. (1996) Technology foresight: Capturing the benefits from science-related technologies. *Research Evaluation* 6 (2), 158.

Martin, B.R. (2010) The origins of the concept of "foresight" in science and technology: An insider's perspective. *Technological Forecasting and Social Change* 77 (9), 1438–1447.

Masini, E.B. (1989) The future of futures studies: A European view. *Futures* 21 (2), 152–160. https://doi.org/10.1016/0016-3287(89)90003-7.

Masini, E. (2006) Rethinking futures studies. *Futures* 38 (10). https://doi.org/10.1016/j.futures.2006.02.004.

Mason, A., Lee, R. and Network, N. (2022) Six ways population change will affect the global economy. *Population and Development Review* 48 (1), 51–73.

Matunga, H., Matunga, H.P. and Urlich, S. (2020) From exploitative to regenerative tourism: Tino rangatiratanga and tourism in Aotearoa New Zealand.

MBIE (2023) *He Mahere Tiaka Kaimahi - Better Work Action Plan*. Wellington: Ministry of Business, Innovation and Employment. See https://www.mbie.govt.nz/immigration-and-tourism/tourism/tourism-projects/tourism-industry-transformation-plan/phase-1-better-work-he-mahere-tiaki-kaimahi/he-mahere-tiaki-kaimahi-better-work-action-plan/.

McDonald, P. (2011) It's time for management version 2.0: Six forces redefining the future of modern management. *Futures* 43 (8), 797–808. https://doi.org/10.1016/j.futures.2011.05.001.

Meadows, D.H., Meadows, D.L, Randers, J. and Behrens III, W.W. (1972) *The Limits to Growth; A Report for the Club of Rome's Project on the Predicament of Mankind*. Universe Books.

Mendelow, A. (1991) Stakeholder mapping. *Proceedings of the Second International Conference on Information Systems* (pp. 10–24). Éditeur inconnu.

Mika, J.P. and Scheyvens, R.A. (2022) Te Awa Tupua: Peace, justice and sustainability through Indigenous tourism. *Journal of Sustainable Tourism* 30 (2–3), 637–57.

Mingers, J. (2014) *Systems Thinking, Critical Realism and Philosophy: A Confluence of Ideas*. Routledge. http://vuw.eblib.com/patron/FullRecord.aspx?p=1682233.

Mingers, J. and Brocklesby, J. (1997) Multimethodology: Towards a framework for mixing methodologies. *Omega* 25 (5), 489–509. https://doi.org/10.1016/S0305-0483(97)00018-2.

Mingers, J. and Rosenhead, J. (2004) Problem structuring methods in action. *European Journal of Operational Research* 152 (3), 530–554. https://doi.org/10.1016/S0377-2217(03)00056-0.

Mitchell, D.K., Agle, B.R. and Wood, D.J. (1997) Toward a theory of stakeholder identification and salience: Defining the principle of who and what really counts. *The Academy of Management Review* 22 (4), 853–886. https://doi.org/10.2307/259247.

Moriarty, J. (2012) Theorising scenario analysis to improve future perspective planning in tourism. *Journal of Sustainable Tourism* 20 (6), 779–800. https://doi.org/10.1080/09669582.2012.673619.

Munro, C. and Yeoman, I.S. (2005) Impact of the macro environment: An examination of the economic propensity of UK regional markets for tourism to Scotland. *Journal of Vacation Marketing* 11 (4), 370–381.

My London News (7 August 2015) In pictures: Notting Hill carnival through the years. See https://www.mylondon.news/whats-on/music-nightlife-news/pictures-notting-hill-carnival-through-9813806 (accessed September 2021).

NBTC Holland Marketing (2008) *Destinatie Holland 2020. Toekomstvisie inkomend toerisme*. NBTC Holland Marketing.

NBTC Holland Marketing (2013a) *Destinatie Holland 2025. Toekomstvisie inkomend toerisme*. NBTC Holland Marketing.

NBTC Holland Marketing (2013b) *Holland Branding & Marketing Strategie 2018*. NBTC Holland Marketing.

NBTC Holland Marketing (2015) *Holland Branding & Marketing Strategie 2020*. NBTC Holland Marketing.

NBTC Holland Marketing (2019) *Perspectief 2030. Bestemming Nederland*. NBTC Holland Marketing.

Nekkers, J. (2009, 3e dr.) *Wijzer in de toekomst. Werken met toekomstscenario's*. Business Contact.

OECD (2019) Strategic foresight for better policies. building effective governance in the face of uncertain futures. https://www.oecd.org/strategic-foresight/ourwork/Strategic%20Foresight%20for%20Better%20Policies.pdf (accessed April 2022).

Oner, A.M. (2010) On theory building in foresight and futures studies: A discussion note. *Futures* 42 (October), 12. http://www.sciencedirect.com.helicon.vuw.ac.nz/science/journal/00163287/42/9.

Ozbekhan, H. (1977) The future of Paris: A systems study in strategic urban planning. *Philosophical Transactions of the Royal Society London, Series A, Mathematical and Physical Sciences* 287 (1346), 523–544.

Page, S. (2022) *Demographic composition of the tourism workforce*. Presentation to The Tourism Industry Transformation Plan (ITP) Group. Tourism. Ministry of Business, Innovation and Employment. Wellington.

Page, S., Yeoman, I.S., Munro, C., Connell, J. and Walker, L. (2006) A case study of best practice – Visit Scotland's prepared response to an influenza pandemic. *Tourism Management* 27 (3), 361–393. https://doi.org/10.1016/j.tourman.2006.01.001.

Pawson, R., Wong, G. and Owen, L. (2011) Known knowns, known unknowns, unknown unknowns: The predicament of evidence-based policy. *American Journal of Evaluation* 32 (4), 518–546.

Pearce, D.G. (2012) *Frameworks for Tourism Research*. CABI.

Peeters, P. and Papp, B. (2023) *Envisioning Tourism in 2030 and Beyond. The Changing Shape of Tourism in a Decarbonising World*. Travel Foundation. https://s3-eu-west-1.amazonaws.com/travelfoundation/wp-content/uploads/2023/02/28113213/EnvisionTourism_Full_FINAL.pdf.

Peeters, P., Gössling, S., Klijs, J., Milano, C., Novelli, M., Dijkmans, C., Eijgelaar, E., Hartman, S., Heslinga, J., Isaac, R., Mitas, O., Moretti, S., Nawijn, J., Papp, B. and Postma, A. (2018) *Research for TRAN Committee – Overtourism: Impact and Possible Policy Responses*. Brussels: European Parliament, Policy Department for Structural and Cohesion Policies. https://www.cstt.nl/Projects/Research-study-on-Overtourism--impact-and-possible-policy-responses/30).

Penn, M. (2007) *Microtrends: The Small Forces Behind Tomorrow's Big Changes*. Hachette.

Penn, M. and Fineman, M. (2018) *Microtrends Squared: The New Small Forces Driving Today's Big Disruptions*. Simon and Schuster.

Peters, G. and Woolley, J.T. (n.d.) Herbert Hoover, statement on the report of the-president's research committee on social trends, 2 January 1933. *The American Presidency Project*. See https://www.presidency.ucsb.edu/node/207918 (accessed March 2022).

Piaget, J., and Brown, T. (1985) *The Equilibration of Cognitive Structures: The Central Problem of Intellectual Development*. University of Chicago.

Pine, R. and McKercher, B. (2004) The impact of SARS on Hong Kong's tourism industry. International *Journal of Contemporary Hospitality Management* 16 (2), 139–143.

Porter, M.E. (1980) *Competitive Strategy*. Free Press.

Porter, M.E. (1985) *Competitive Advantage*. Free Press.

Porter, M.E. (1990) *The Competitive Advantage of Nations*. Free Press.

Poskitt, S., Waylen, K.A. and Ainslie, A. (2021) Applying pedagogical theories to understand learning in participatory scenario planning. *Futures* 128, 102710.

Postma, A. (2013a) Anticipating the future of European tourism. In A. Postma, I. Yeoman and J. Oskam (eds) *The Future of European Tourism* (pp. 290–305). European Tourism Futures Institute.

Postma, A. (2013b) When the Tourists Flew In. Critical Encounters in the Development of Tourism. Unpublished draft PhD thesis, University of Groningen, Faculty of Spatial Sciences.

Postma, A. (2015a) Overuse of city-destinations: Limits and solutions. Keynote *ITB – Futures Day*. Berlin, 5 March 2015. Co-referent: Lars Bernhard Jørgensen, CEO Wonderful Copenhagen.

Postma, A. (2015b) Investigating scenario planning – A European tourism perspective. *Journal of Tourism Futures* 1 (1), 46–52. http://dx.doi.org/10.1108/JTF-12-2014-0020.

Postma, A. and Papp, B. (2021) Of trends and trend pyramids. *Journal of Tourism Futures* 7 (2), 162–167. https://doi.org/10.1108/JTF-11-2019-0129.

Postma, A. and Schmücker, D. (2017) Understanding and overcoming negative impacts of tourism in city destinations: Conceptual model and strategic framework. Special Issue: The Future of City Tourism. *Journal of Tourism Futures* 3 (2), 144–156. https://doi.org/10.1108/JTF-04-2017-0022.

Postma, A. and Yeoman, I.S. (2016) Conceptualising scenario planning in tourism futures: Vignettes of practice. *CAUTHE 2016: The Changing Landscape of Tourism and Hospitality: The Impact of Emerging Markets and Emerging Destinations*. Sydney: Blue Mountains International Hotel School. https://search.informit.org/documentSummary;dn=936497265387202;res=IELBUS.

Postma, A. and Yeoman, I.S. (2021) A systems perspective as a tool to understand disruption in travel and tourism. *Journal of Tourism Futures* 7 (1), 67–77. https://doi.org/10.1108/JTF-04-2020-0052.

Postma, A., Heslinga, J., Hartman, S. (2020a) *Four Futures Perspectives of the Visitor Economy After CoVid-19*. Centre of Expertise in Leisure, Tourism and Hospitality.

Postma, A., Koens, K. and Papp, B. (2020b) Overtourism: Carrying capacity revisited. In J. Oskam (ed.) *The Overtourism Debate: NIMBY, Nuisance, Commodification* (pp. 149–159). Emerald Publishers. https://doi.org/10.1108/978-1-83867-487-820201015.

Postma, A., Papp, B. and Koens, K. (2018) *Visitor Pressure and Events in an Urban Setting: Understanding and Managing Visitor Pressure in Seven European Urban Tourism Destinations*. Breda/Leeuwarden/Vlissingen: CELTH. https://www.etfi.eu.

Powell, T.C. (2001) Fallibilism and organizational research: The third epistemology. *Journal of Management Research* 1 (4), 19. http://search.proquest.com.helicon.vuw.ac.nz/abiglobal/docview/237239511/fulltextPDF/126CF5F2AA014D70PQ/42?accountid=14782.

Provincie Friesland (2019) *Geluk op 1. Vernieuwen in Vertrouwen. Bestuursakkoord 2019–2023* (Leeuwarden: Provincie Friesland).

Quorin, M., Eeckhout, L. and Harrison, D. (2020) The new next: Which consumer trends have been boosted by the pandemic. Foresisight Factory. https://www.foresightfactory.co/webinar-on-demand/#boosted (accessed January 2024).

Raäisaänen, J. (2007) How reliable are climate models? *Tellus A: Dynamic Meteorology and Oceanography* 59 (1), 2–29.

Ramírez, R. and Ravetz, J. (2011) Feral futures: Zen and aesthetics. *Futures* 43 (4), 478–487. https://doi.org/10.1016/j.futures.2010.12.005.

Ramirez, R. and Wilkinson, A. (2014a) Rethinking the 2×2 scenario method: Grid or frames? *Technological Forecasting and Social Change* 86, 254–264. https://doi.org/10.1016/j.techfore.2013.10.020.

Ramirez, R. and Wilkinson, A. (2014b) Rethinking the 2×2 scenario method: Grid or frames? *Technological Forecasting and Social Change* 86, 254–264.

RAND (2022) *A Brief History of RAND* (n.d.) https://www.rand.org/about/history.html (accessed March 2022).

Ransfield, A.K. and Reichenberger, I. (2021) Māori Indigenous values and tourism business sustainability. *AlterNative: An International Journal of Indigenous Peoples* 17 (1), 49–60.

Roberts, C. (2022) Re-evaluating New Zealand tourism–what the future holds. *Journal of Tourism Futures* (ahead-of-print). https://doi.org/10.1108/JTF-01-2022-0026.

Rowland, N.J. and Spaniol, M.J. (2017) Social foundation of scenario planning. *Technological Forecasting and Social Change* 124, 6–15. https://doi.org/10.1016/j.techfore.2017.02.013.

Sandberg, A. (1978) *The Limits to Democratic Planning.* LiberForlag. https://doi.org/10.2307/3009284.

Sardar, Z. (2010) The Namesake: Futures; futures studies; futurology; futuristic; foresight—What's in a name? *Futures* 42 (3), 177–184. https://doi.org/10.1016/j.futures.2009.11.001.

Saritas, O. and Smith, J.E. (2011) The Big Picture – Trends, drivers, wild cards, discontinuities and weak signals. *Futures* 43 (3), 292–312. https://doi.org/10.1016/j.futures.2010.11.007.

Schänzel, H.A. and Smith, K.A. (2011) The absence of fatherhood: Achieving true gender scholarship in family tourism research. *Annals of Leisure Research* 14 (2–3), 143–154. https://doi.org/10.1080/11745398.2011.615712.

Scheyvens, R., Carr, A., Movono, A., Hughes, E., Higgins-Desbiolles, F. and Mika, J.P. (2021) Indigenous tourism and the sustainable development goals. *Annals of Tourism Research* 90, 103260. https://doi.org/10.1016/j.annals.2021.103260.

Schnaars, S.P. (1987) How to develop and use scenarios. *Long Range Planning* 20 (1), 105–114. https://doi.org/10.1016/0024-6301(87)90038-0.

Schoemaker, P.J.H. (1993) Multiple scenario development: Its conceptual and behavioral foundation. *Strategic Management Journal.* 14 (3), 193–213. https://doi.org/10.1002/smj.4250140304.

Schwartz, P. (1991) *The Art of the Long View. Planning for the Future in an Uncertain World.* Currency Doubleday.

Schwartz, P. (1996) *The Art of the Long View: Paths to Strategic Insight for Yourself and Your Company.* Prospect Publishing.

Sellberg, M.M., Wilkinson, C. and Peterson, G.D. (2015) Resilience assessment: A useful approach to navigate urban sustainability challenges. *Ecology and Society* 20 (1), 43. http://dx.doi.org/10.5751/ES-07258-200143.

Seyitoğlu, F. and Costa, C. (2022) A systematic review of scenario planning studies in tourism and hospitality research. *Journal of Policy Research in Tourism, Leisure and Events* 1–18. https://doi.org/10.1080/19407963.2022.2032108.

Sheptycki, J. (2020) The politics of policing a pandemic panic. *Australian and New Zealand Journal of Criminology* 53 (2), 157–173.

Singh, S. (2020) "Quixotic" tourism? Safety, ease, and heritage in post-COVID world tourism. *Journal of Heritage Tourism* 16 (6), 716–721. https://doi.org/10.1080/1743873X.2020.1835924.

Slaughter, R.A. (1996) Foresight beyond strategy: Social initiatives by business and government. *Long Range Planning* 29 (2), 156–163. https://doi.org/10.1016/0024-6301(96)00003-9.

Slaughter, R.A. (2002) From forecasting and scenarios to social construction: Changing methodological paradigms in futures studies. *Foresight* 4 (3), 26–31. https://doi.org/10.1108/14636680210697731.

Smith, S. and Lee, H. (2010) A typology of 'theory' in tourism. In D. Pearce and R. Butler (eds) *Tourism Research: A 20-20 Vision* (pp. 28–39). Goodfellows Publishers.

Spaniol, M.J. and Rowland, N.J. (2018) The scenario planning paradox. *Futures* 95, 33–43. https://doi.org/10.1016/j.futures.2017.09.006.

Taleb, N.N. (1997) *Dynamic Hedging: Managing Vanilla and Exotic Options* 64. John Wiley and Sons.

Thompson, M. (2011) Ontological shift or ontological drift? Reality claims, epistemological frameworks and theory generation in organisational studies. *Academy of Management Review* 36 (4), 20. https://doi.org/10.5465/amr.2010.0070.

Towner, N. and Lemarié, J. (2020) Localism at New Zealand surfing destinations: Durkheim and the social structure of communities. *Journal of Sport and Tourism* 24 (2), 93–110. https://doi.org/10.1080/14775085.2020.1777186.

UNPAN (2021) CEPA strategy guidance note on strategic planning and foresight. United Nations Public Administration network (UNPAN). See https://unpan.un.org/sites/unpan.un.org/files/Strategy%20note%20%20strategic%20foresight%20Mar%20 2021_1.pdf (accessed April 2023).

UNWTO (2018) *"Overtourism"? Understanding and Managing Urban Tourism Growth Beyond Perceptions Volume 1.* UNWTO. https://doi.org/10.18111/9789284419999.

van 't Klooster, S. A. and van Asselt, M.B.A. (2006) Practising the scenario-axes technique. *Futures* 38 (1), 15–30. https://doi.org/https://doi.org/10.1016/j.futures.2005.04.019.

Van der Heijden, K. (1996, 2004) *Scenarios: The Art of Strategic Conversation*. John Wiley.

Van der Heijden, K. (2005) *Scenarios: The Art of Strategic Conversation* (2nd edn). John Wiley. https://doi.org/http://www.loc.gov/catdir/toc/ecip0421/2004018710.html.

Van der Heijden, K., Bradfield, R., Burt, G., Cairns, G. and Wright, G. (2002) *Sixth Sense Accelerating Organizational Learning with Scenarios*. Wiley.

van Zon, H. (1992) Alternative scenarios for Central Europe. *Futures* 24 (5), 471–482. https://doi.org/10.1016/0016-3287(92)90017-A.

Varum, C.A. and Melo, C. (2010) Directions in scenario planning literature—A review of the past decades. *Futures* 42 (4), 355–369. https://doi.org/10.1016/j.futures.2009.11.021.

Verganti, R. (2009) *Design Driven Innovation: Changing the Rules of Competition by Radically Innovating What Things Mean*. Harvard Business Press.

Voros, J. (2017) The futures cone, use and history. https://thevoroscope.com/2017/02/24/the-futures-cone-use-and-history/ (accessed December 2021).

Wack, P.A. (1984) *Scenarios: The Gentle Art of Re-perceiving: One Thing or Two Learned while Developing Planning Scenarios for Royal Dutch/Shell*. Division of Research, Harvard Business School.

Wack, P. (1985) Scenarios: Uncharted waters ahead. *Harvard Business Review* 63 (5), 72.

Wade, W. (2021) *See Your New Normal: A How-To Guide to Excelling in Your Post-COVID Future Using Scenario Planning*. Wade and Company SA.

Walker, B., Holling, C.S., Carpenter, S.R. and Kinzig, A. (2004) Resilience, adaptability and transformability in social–ecological systems. *Ecology and Society* 9 (2), 5. http://www.ecologyandsociety.org/vol9/iss2/art5.

Weick, K.E. and Roberts, K.H. (1993) Collective mind in organizations: Heedful interrelating on flight decks. *Administrative Science Quarterly* 38 (3), 357–381. https://doi.org/10.2307/2393372.

Wells, H.G. (1902) The discovery of the future. *Nature* 65 (1684), 326–331.

Wells, H.G. (1977) *The Time Machine: The War of the Worlds* (A critical edn./edited by Frank D. McConnell). Oxford University Press.

Wilkinson, A., Kupers, R. and Mangalagiu, D. (2013) How plausibility-based scenario practices are grappling with complexity to appreciate and address 21st century challenges. *Technological Forecasting and Social Change* 80 (4), 699.

Wilson, I. (2000) From scenario thinking to strategic action. *Technological Forecasting and Social Change* 65 (1), 23–29. https://doi.org/10.1016/S0040-1625(99)00122-5.

Wolfram, G. and Burnill-Maier, C. (2013) The tactical tourist. In M. Smith and G. Richards (eds) *The Routledge Handbook of Cultural Tourism* (pp. 361–368). Routledge. https://pure.uvt.nl/ws/portalfiles/portal/93704006/Handbook_of_Cultural_Tourism_Full_Text_Revised.pdf.

Yeoman, I.S. (2004) The development of a conceptual map of soft operational research practice. Unpublished thesis, Edinburgh Napier University. http://researchrepository.napier.ac.uk/id/eprint/3873.

Yeoman, I.S. (2012a) *2050 Tomorrow's tourist*. https://www.tomorrowstourist.com/index.php (accessed July 2022).

Yeoman, I.S. (2012b) Authentic learning: My reflective journey with postgraduates. *Journal of Teaching in Travel and Tourism* 12 (3), 295–311. https://doi.org/10.1080/15313220.2012.704258.

Yeoman, I.S. (2016) The Future Tourist: Fluid and Simple Identities (30/06/2016: Victoria University of Wellington). https://vimeo.com/181103735/ad143522da.

Yeoman, I.S. and Mars, M. (2012) Robots, men and sex tourism. *Futures* 44 (4), 365–371. https://doi.org/10.1016/j.futures.2011.11.004.

Yeoman, I.S. and McMahon-Beattie, U. (2005) Developing a scenario planning process using a blank piece of paper. *Tourism and Hospitality Research* 5 (3), 273–285. https://doi.org/10.1057/palgrave.thr.6040026.

Yeoman, I.S. and McMahon-Beattie, U. (2014) New Zealand tourism: Which direction would it take? *Tourism Recreation Research* 39 (3), 415–435. https://doi.org/10.1080/02508281.2014.11087009.

Yeoman, I.S. and McMahon-Beatte, U. (2016) An ontological classification of tourism futures. *CAUTHE 2016: The Changing Landscape of Tourism and Hospitality: The Impact of Emerging Markets and Emerging Destinations*. Sydney. http://search.informit.com.au/documentSummarydn=900181604404970res=IELBUS.

Yeoman, I.S. and McMahon-Beatte, U. (2018a) The future of luxury: Mega drivers, new faces and scenarios. *Journal of Revenue and Pricing Management* 17, 204–217. https://doi.org/10.1057/s41272-018-0140-6.

Yeoman, I.S. and McMahon-Beatte, U. (2018b) Framing tourism futures research: An ontological perspective. In C. Cooper, S. Volo, W.C. Gartner and N. Scott (eds) *The SAGE Handbook of Tourism Management: Applications of Theories and Concepts to Tourism* (pp. 463–77). Sage.

Yeoman, I. and McMahon-Beattie, U. (eds) (2020) *The Future Past of Tourism: Historical Perspectives and Future Evolutions*. Channel View Publications.

Yeoman, I.S. and Postma, A. (2014) Developing an ontological framework for tourism futures. *Tourism Recreation Research* 39 (3), 299–304. https://doi.org/10.1080/02508281.2014.11087002.

Yeoman, I.S., Galt, M. and McMahon-Beattie, U. (2005a) A case study of how VisitScotland prepared for war. *Journal of Travel Research* 44 (1), 6–20.

Yeoman, I.S., Lennon, J.J. and Black, L. (2005b) Foot-and-mouth disease: A scenario of reoccurrence for Scotland's tourism industry. *Journal of Vacation Marketing* 11 (2), 179–190. https://doi.org/10.1177/1356766705052574.

Yeoman, I., McMahon-Beatte, U. and Sigala, M. (eds) (2021) *Science Fiction, Disruption and Tourism*. Channel View Publications.

Yeoman, I.S., Postma, A. and Hartman, S. (2022) Scenarios for New Zealand tourism: A COVID-19 response. *Journal of Tourism Futures* 8 (2), 177–193. https://doi.org/10.1108/JTF-07-2021-0180.

Yeoman, I.S., Postma, A. and Oskam, J. (2015a) Editorial. *Journal of Tourism Futures* 1 (1), 3–5. https://doi.org/10.1108/JTF-12-2014-0016.

Yeoman, I.S., Schänzel, H.A. and Zentveld, E. (2022a) Family tourism: A New Zealand COVID-19 perspective. *Journal of Tourism Futures* (ahead-of-print). https://doi.org/10.1108/JTF-12-2021-0274.

Yeoman, I.S., Schzanel, H. and Zentveld, E. (2022b) Tourism trends in a COVID-19 world: A New Zealand perspective. *Journal of Tourism Futures* (published online in advance of print).

Yeoman, I.S., Andrade, A., Leguma, E., Wolf, N., Ezra, P., Tan, R. and McMahon-Beattie, U. (2015b) 2050: New Zealand's sustainable future. *Journal of Tourism Futures* 1 (2), 117–130. https://doi.org/10.1108/JTF-12-2014-0003.

Zheng, D., Luo, Q. and Ritchie, B.W. (2021) Afraid to travel after COVID-19? Self-protection, coping and resilience against pandemic "travel fear". *Tourism Management* 83, 104261. https://doi.org/10.1016/j.tourman.2020.104261.

Index

Academy of Leisure and Tourism xv, 3
adaptation 21, 27–9, 95, 99, 136, 160, 212, 219
adaptive, adaptive power xiii, 17, 35, 38, 59, 128–9, 192, 202, 219
adaptive capacities xiii, 22–3, 28, 96, 186, 203, 216
agriculture 104, 106–8, 110–12, 114–15, 117–19, 121–3, 132, 136–7, 143
Airbnb 61, 152, 154, 187, 201
ambiguity 33, 57
Amsterdam 7, 26, 149, 151–4, 186–8, 194, 211
anticipation 37–8
ANVR xv, 211
Appreciative Enquiry 131, 165–6
arena approach 34
artificial intelligence (AI) 76
attitude/s xiii, 31–2, 36, 88, 94, 149, 152, 156, 160, 166, 186, 198, 210, 214
  passive attitude 36
  pre-active attitude 36
  pro-active attitude 36, 148
  reactive attitude 36

bachelor level xvi
Bandhold, Hans 32, 38
Berger, Gaston 20, 35
Bergman, Anne 14, 16–17, 218
Bergman's Future States 218
best-case scenario 29
biodiversity 24, 107, 111, 115, 118, 121, 123, 136, 207
Bishop, Peter 10–11, 60–2, 83, 208
black swans 45, 130
bottom-up approach 165
bounded rationality 33
Bournemouth University 163, 183, 209
brainstorming 50, 81
brand awareness/identity 182, 208, 210
BSR-lifestyle model 198
Burnill-Maier, Claire 187
business model 25, 39, 63, 78, 82, 165
Butler, Richard 207

Carnival community, Carnivalists 6, 162–7, 169, 182, 213
case studies 5–6
causal layered analysis, CLA 32, 67
cause-effect relationships, causality 67–8
Cavagnaro, Elena xv
Centre of Expertise in Leisure, Tourism and Hospitality (CELTH) 7, 186, 210
certificates of participation 208
challenges 5, 17, 29, 38, 43, 63, 100, 122, 128–9, 131, 144, 149, 163–5, 207, 213, 216, 219
change management 166
ChatGPT 76
circular economy 95, 122, 217
climate change xvii, 2, 5, 9–10, 24, 41, 58, 99, 102, 107, 110, 113–14, 117–18, 121–2, 128, 136, 143, 156, 159–60, 187, 207, 217
co-creation 130
coastal defence 106, 110, 114, 118, 122
collaboration xviii, 7, 50, 77, 131, 133, 155, 181, 208, 219
collaborative understanding 68
collaborative scenario planning, collaborative planning 163, 219
collaborative workshops 131
communities of practice 19–20
community 43, 94–5, 100, 146, 166, 168, 171–2, 196–7, 219
competencies 38, 210
complex adaptive systems 23, 27, 64, 83
complexity 10, 12, 14, 18, 20–1, 25, 33, 36, 41, 45, 58, 92, 129–30
conditions 19, 24, 28, 34, 56, 61, 79, 98, 109–10, 148, 151, 168, 191, 203, 215, 218
  context 5–6, 9, 12, 18, 25, 27, 36, 38, 60, 89, 91, 99, 129, 144, 151, 156, 171, 188, 198, 215–19
  organisational 11, 14, 25, 28, 38, 58, 93, 145, 214–15

233

234 Index

conceptualisation 218
conscious destination 153–4
consensus 14, 34, 38, 64, 67, 87, 90, 168
consolidation 34, 114, 144–5, 213
Conway, Maree 9, 36, 129, 215
Cooperrider, David L. 167
coping strategies 203
countryside 109, 114, 121–3, 127, 211
COVID-19 1, 5–6, 10, 24, 36, 41, 43, 45, 47–9, 53–4, 60, 64, 83–7, 89, 92, 94, 98, 127, 132, 145, 207, 210, 212–13
creative space 131
creativity 17, 38, 41, 50, 52, 58, 69, 92–4, 124, 129, 168, 180, 192
crisis 17, 24, 36, 41, 45–6, 86–92,
cultural understanding 189, 192

data driven society 194
Dator, James 12, 16–17, 21, 55, 86, 218
De Geus, Arie 35
De Jouvenel, Hugues 35
decision making xvii–xviii, 10, 15, 18–19, 21, 33, 35–6, 76, 105, 214–15, 218
design 28, 41, 50–1, 58, 78, 83, 100–1, 104, 122, 131, 167, 213, 215–16
desirable future 39, 42
destination development 25–6, 85
destination governance 216
destination leadership 45–6
Destination Management Organisation (DMO) 7, 26, 41, 147, 186, 188, 190, 194, 212
destination resilience 22
dialogue/s 6, 15, 18, 21, 120, 124, 128–9, 131, 156–7
diaspora 169
dilemma/s 72, 87, 89–90, 92, 105, 122–3
disaster 24, 219
discontinuities 155
discontinuous event, discontinuous force 60, 62
dislikes 81, 144
disruptions 10, 24, 40, 47, 62, 92, 130, 147 155, 207
diversity 5, 12, 65, 110, 118, 121, 133, 174, 192, 202, 217
domain analysis 148–9, 156
domain map 60, 82, 149–50
Drenthe xv 208
driving forces of change 64, 88, 135, 137, 189–90, 195
Durance, Philippe 11, 36–7, 39, 73, 84

emerging issues 133
Emmen xvi, 208, 211

energy transition 100, 106, 132
entrepreneurial innovations 7, 164, 181–2
environment xiii, 2, 11, 17–18, 20, 36–9, 47, 51, 58–9, 62, 70, 72, 80, 88, 90, 95–6, 98, 100, 122, 125, 127, 135, 138–9, 144, 152, 158, 161, 176, 178, 180, 182, 186, 188, 192, 194, 197, 202, 212–14, 217, 219
  business environment 14, 17, 35, 37, 57–8, 72
  contextual environment 27, 63, 67, 133
  external environment xiii, 68, 130
  macro-environment 65, 133–4, 194, 214
  meso-environment 65, 67, 72, 80, 133, 144
  micro-environment 65
  transactional environment 134
environmental analysis 64, 66–7
environmental impact 149, 194
epistemology 16, 218
  constructivist epistemology 33
  positivist epistemology 32
era analysis 60–1, 148, 151, 153
European cities xiii, 4, 7, 186, 188–90, 194, 202, 210–12, 214
  Amsterdam 7, 26, 149, 151–4, 186–8, 194, 211
  Antwerp 7, 188, 194, 200, 202, 211
  Barcelona 7, 186, 188, 194, 211
  Berlin 4, 7, 162, 186–8, 194, 208–9, 211
  Bruges 7, 188, 194, 200, 211
  Copenhagen 7, 187–8, 194, 209, 211
  Ghent 7, 188, 194, 200, 211
  Leuven 7, 188, 194, 200, 211
  Lisbon 7, 186, 188, 194, 209, 211
  Mechelen 7, 188, 194, 200, 211
  Munich 7, 186, 188, 194, 211
  Riga 186
  Salzburg 7, 187–8, 194, 200, 211
  Tallinn 7, 187–8, 194, 200, 211
European Foresight Platform 37, 129–30
ETFI/European Tourism Futures Institute xiii, xv–xvii, 1, 3–6, 18–19, 21, 40–1, 58–9, 62–6, 70, 73, 75–84, 86–7, 100–1, 104, 129, 133, 144–5, 148, 155, 163–5, 180–1, 186, 188, 194, 203, 207–8, 210–17
European Tour Operator Association (ETOA) 4, 209–10
evaluation 3, 47, 50, 58, 63, 140, 183–4, 203, 216
  ex ante evaluation 64, 78, 82
  ex poste evaluation 78

evolutionary perspective 26
evolution 30, 207
exercise 49, 73, 81, 83, 133, 200
  group exercise 167, 169
  individual exercise 167
expected future 149, 154–5, 157
experts 7, 32, 62, 83, 86–8, 130–1, 165, 171, 173–4, 188, 213
experience worlds 198–9
explanatory claims 16–17
exponential growth, exponential decline 45, 130
external business environment 35
extrapolation 12, 32

facilitation 20, 131–2, 152
fast variables 24
Ferdinand, Nicole 162–3, 184
financial crisis xv, 10, 44–5
financing model 174
forces of change 58, 64, 88, 99, 135, 137, 189–90, 195
forecasting 9, 13, 31, 33, 51, 161, 211
forecasting model 161
foresight, strategic foresight xiii, xvi, 1–6, 8–9, 12, 16, 21–2, 30, 35–40, 42, 45, 58–9, 62–4, 83–4, 125, 129–30, 133, 135, 147, 149, 164, 181, 187–8, 207, 210, 214–17
force field 86, 96–7, 128, 135, 145, 155, 157
frameworks 13, 15, 25, 42, 98, 122–3, 218
Friesland 6, 99–102, 104–6, 109–10, 113–14, 117, 121–2, 125–8, 134, 145, 211–13
funding 163, 174–6, 178–80, 182–3
future xiv, xvi–xviii, 1, 3–10, 12–18, 20–1, 26, 28, 30–43, 45, 47, 50–5, 57–62, 70, 75–6, 78, 80–2, 84–6, 88–91, 96–101, 104–5, 121–5, 128, 132, 136, 138–40, 144, 149, 151, 154–7, 163–6, 168, 171, 173–4, 176, 178–80, 182, 184, 186–8, 190, 194, 199–200, 202–3, 205, 207–19
  preferred future 3, 59
  probable future 59, 62
  possible future 22, 28, 59
  plausible future 96, 218
  single future 30, 40, 60, 213
  surprise free future 32, 61
future of work 41, 53–4, 56, 58
future proof 1, 3, 5, 30, 35, 37–8, 84, 86, 130, 145, 182, 188, 200, 202, 212–15
future scanning 64–5, 132–5
future states 10, 22, 25, 28, 218
Futures Lab 6, 125, 132–3, 145, 212–13

futures studies 12–14, 16, 21, 33, 209–10
futures thinking 30, 33, 40, 207, 216
futures wheel 79–80, 144
futurism, futurology 33

Gidley, Jennifer 30–1
Godet, Michel 11, 36–7, 39, 73, 84
governance xiii, 22, 27–8, 42, 45, 100, 107, 109, 111, 113–14, 116, 118, 120, 137, 152, 187, 191, 206, 212, 216
  adaptive governance 129
  shared governance 129
governance framework 123
governance implications 28, 206
government xiii, xv, xviii, 1, 6, 9, 31, 39, 41–2, 45, 54, 87–90, 92–4, 96–7, 100, 108, 112, 116, 119, 121–3, 125–6, 129, 131, 133, 141, 152, 154, 182, 191–2, 196–8, 219

Hartman, Stefan xiii, 2–3, 23, 25, 27–8, 78, 129, 206, 216
higher education xv
Hines, Andy 60–2, 83, 208
history xv, 21, 117, 152, 168, 175–6, 179, 218
HISWA-RECRON 212
horizon scanning 63–7, 132–3, 155, 188–9, 194
hybrid workshops 84

identity 22, 27, 49, 61, 121–3, 125, 128, 143, 151, 174–6, 182, 196–8, 208
images, visualisations 34, 59, 62, 76–7, 104, 139
imagination 15, 38, 69, 74, 139
implication tree 80–1, 144
implications xvii, 6, 19, 28, 45–6, 54–5, 62, 80–1, 90, 99, 101, 104, 124, 132, 140, 143–4, 148–9, 170, 182, 187–8, 191–3, 200, 202, 205–6, 212
  direct, first order 80–1, 144, 202
  indirect, second order 81, 144, 202
importance by uncertainty matrix x, 64, 69, 71, 72, 75, 103, 136
Inayatullah, Sohail 67
industry associations xv, 133, 208, 214
informed decisions 218–19
institutional frameworks 25
IT 194
ITB (Internationale Tourismusbörse Berlin) 4, 66, 186, 208–9

Journal of Tourism Futures xiii–xiv, 4, 12, 208, 210

key certainties 64, 75, 136, 138–9, 143
key uncertainties, critical uncertainties 64, 70, 72, 87, 89, 91, 102–3, 136–9, 143, 147, 149, 155–6, 164–5, 171, 174, 189–90, 194–5, 200
King's College London 163, 184
knowledge development xv, 33, 130, 214–16
knowledge economy xv
Koens, Ko 187–8, 203–4, 210

landscape 104, 106–7, 109–11, 114–15, 117–19, 121–3, 125, 143, 159–60, 211, 218
large-scale 106, 108, 112, 114, 116, 120–1
learning frameworks 8, 17
Leeuwarden xiii, xv, 40, 102, 125, 134, 208–9, 211–12
lifestyles, leisure lifestyles 52, 198–201
likes 81, 144
limits of the plausible 64, 69, 189
Lindgren, Mats 32, 38
linear development 32, 61
local authority 121, 123
London 162–3, 172, 174, 178–9, 184–5, 209, 211, 213

management strategies 203–4, 219
master classes 208
master level xvi
McMahon-Beattie, Una 9, 13, 16, 20, 40, 45, 55, 57, 91, 218
mega-event/s 162, 179
Mendelow, Aubrey 61
mental models 11, 15, 18–21
methodologies 5, 9, 11, 23, 26, 28–30, 216–17
Miller, Kenneth xv
mitigation 99, 102, 160
moderated sessions 59, 64
Moffat model 40
monitoring 3, 46, 95, 106, 110, 133, 214
multicultural event/s 162
multiple futures 5, 9, 12, 30, 32–3, 35, 213, 218

nature conservation 217
NBTC Holland Marketing 6, 62, 147–50, 161, 211, 213
Nekkers, Jan 83, 208, 215
Netherlands xiii, xv–xvi, 4, 6, 40–1, 62, 79, 83, 99, 102, 125–9, 147–55, 158, 161, 206, 208–13, 215
New Media Angels 182
New Zealand xiii, 40–4, 47–9, 54–8, 210
NHL Stenden xiii, xv, 3, 40, 87, 100–1, 129, 206–8, 210, 213, 217
non-linear, non-linearity 34, 130

Notting Hill Carnival 6–7, 162, 164–9, 171–85, 209, 211, 213

Omgevingslab Fryslân 99–100
online, software platforms 125, 145
  MURAL 64, 80, 83, 125, 132–3, 141, 144
  MIRO 55, 83, 125, 145
  MS Teams 83, 133
  MS Whiteboard 125, 145
ontology 14, 16, 218
ontological plurality 55, 218
open policymaking 131
operational plans 63
opportunities xv, xvii, 38–9, 50, 54, 57, 63–4, 81–2, 91, 98, 100, 104, 113, 118, 130, 133, 144, 152, 157, 163–4, 194, 198, 202, 211, 216, 218–19
organisational learning 11, 14, 145
Oskam, Jeroen xvi
outside-in 80, 144
overcrowding 163, 186
overtourism xiii, 7, 26, 186–8, 200, 202–5, 210, 213–14

pandemics 156, 207, 217
Papp, Bernadett 2, 28, 47, 65, 134
paradigm 12–13, 15, 33–4, 36, 38, 40, 78, 155, 218
participants 10, 18, 20, 32, 34, 37, 51–2, 64–5, 67–71, 76, 78, 81, 83, 101–2, 104, 120–2, 124, 129, 131–5, 138–41, 143–5, 162–3, 165–9, 171, 173–4, 179, 180, 184, 189, 208
participatory 10, 15, 18, 21, 35, 37, 129, 162, 184
Pearce, Douglas G. 15, 218
Pearce's Tourism Research Frameworks approach 218
Peeters, Paul 2, 26
people-centred design 131
perceived level of impact 136–7, 189
perceived level of unpredictability/uncertainty 136
philosophy 14, 16–7, 42–3, 166
planning horizon 38
plausibility 17, 20
plausible extremes 89, 155, 189
pluralism 33
policy 31, 34, 38–9, 41–4, 54, 57, 80, 82, 97, 99, 110, 122, 130, 132, 136, 144–5, 148, 191–2, 213, 215, 217
policy cycle 39
policy domains 39, 42, 63, 99, 125, 130, 132, 143–5, 213, 217
policy lab 131

policy options 72, 211
policymaking 6, 42, 85, 100, 131, 189, 191–2, 194
politics 20, 41–3, 58, 163, 213
popular media, media 33, 88, 97, 163, 186
Porter, Michael E. 35
Postma, Albert xiii, 3, 8–9, 11, 16, 28, 40–1, 47, 50, 65, 134, 167, 186–8, 210, 214
prediction 16–17, 30–1, 33, 61, 97, 140, 156, 164, 174
process xiii, 5, 10–11, 14–15, 18–20, 23, 27, 34–5, 37, 39, 45, 50–1, 58, 64, 67–8, 81, 101–2, 128–33, 135, 145, 148, 165, 167–8, 171, 182, 188, 196, 203
    collaborative process 32, 39
    cyclic learning process 38
    group process 15, 19–21, 35
    process oriented 34
    strategic learning process 37, 64, 125, 130
product development xvii, 41, 47, 213
Professional Doctorate 215
pro forma 65–6
prognosis 16–17, 161
projections 9, 55, 61, 72, 153
prospective 35, 37
Province of Fryslân xv, 128, 144, 208, 212
public sector stakeholders 99–100, 122

RAND corporation 31–2
rationality 16, 33, 50, 217
reasoning 21, 217
recreation 73, 94–5, 104–7, 110–11, 114–15, 117–19, 123, 128, 198–9, 211–12, 214
regenerative tourism 26, 41, 54, 94
regional development 6, 99, 101
repeat visitors 169
Research Committee on Social Trends 31
resilience xiii, 5, 22–3, 25–9, 83, 86, 95–6, 129, 149, 156, 160, 206, 213, 215
resilient destinations 22–3, 28
risk management 218
robustness 10, 78, 80, 82, 96, 201, 203

safety 48–9, 106, 110, 122, 152, 154, 156, 163, 172, 174, 176, 193–4, 197, 211
scale 99, 102, 108, 114, 121, 123
scenario/s xiv, xvii, 5–6, 9–10, 14–18, 20, 29, 31–6, 38–41, 44–7, 50–1, 57, 59–60, 62–3, 72, 78, 81, 91–4, 106–11, 113, 116, 118, 132, 137–8, 140–1, 143, 157–60, 164, 176, 178–9, 191–3, 195–8, 200

alternative scenario/s 34, 62–3, 147, 149, 155–7, 161, 216
aspirational scenario/s 38–9, 59, 63, 78, 80
baseline scenario/s 6, 32, 62, 147, 149, 153, 155–7, 161, 212–13, 216
explorative environmental scenario/s 72
explorative scenario/s 30, 32–4, 38–9, 59, 62–4, 70, 72, 79, 85–6, 96, 138, 156, 213
extrapolative scenario/s 32, 59
goal-oriented scenario/s 38, 59
normative scenario/s 38, 78, 216
morphological scenario/s 73
predictive scenario/s 30, 32, 34, 59–62, 72, 96, 213
risk-based scenario/s 216
system scenario/s 72
wild card scenario/s 62
what-if scenario/s 216
scenario cross, scenario matrix 17, 21, 55, 64, 72–5, 138–40, 157, 174, 189
scenario cycle 100
scenario framework 6, 62, 64, 85, 87, 98, 101–3, 140, 174
scenario narrative 75–6, 132, 138
    first person 75
    structured 19, 75, 101–2, 104, 135, 143, 203–4
    unstructured 75
scenario planning xiii, xv–xviii, 1–8, 10–15, 17–24, 26, 28–9, 31–5, 39–48, 50, 53, 55–60, 62–3, 69, 83–5, 98–101, 125, 128–9, 133, 137, 162–3, 165, 180–1, 187–9, 204–5, 207–11, 213–19
scenario question 54, 60, 62, 64, 66, 70, 101, 148
scenario study 6, 149, 151, 161, 163, 188, 210, 213
scenario thinking 10–11, 19, 35, 210, 214
scenario use 20, 55
scepticism 12, 217
Schwartz, Peter 11, 35
science fiction 9, 16–17, 31, 66, 146, 217
sensemaking 14
shocks 24–5, 27, 40, 130
slow variables 24
small-scale 108, 110, 112, 116–20, 122
Smart Agent 198–9
social reality 20
spatial planning xiii, 42, 99–100, 123
SSN Samenwerkingverband Noord-Nederland 208

stakeholders xvii, 3, 19–20, 25, 27–9, 37, 39, 42, 46–7, 54–5, 59–61, 64–5, 75, 80, 82, 85–7, 91, 94, 97–102, 105, 122–3, 125, 129–30, 140, 148, 151, 155, 163–5, 181, 184, 186, 200–1, 203–4, 207–8, 210, 218–19
sticky notes 65, 67–8, 70–1, 80, 82–3, 104
storyline 91, 104, 106, 109, 113, 117, 194, 219
storytelling 75, 106, 110, 114, 118
strategic domains 82
strategic learning 32, 37, 64, 125, 130, 145
strategic thinking/planning xv, 5, 14, 31, 35–6, 63, 166, 193, 214
strategy/ies xiii, 3–6, 10, 14, 18, 30–1, 34–9, 41–2, 45–6, 55, 58, 62, 64, 78, 80, 82, 85, 91–6, 127–9, 131, 143–4, 147–8, 155, 161, 182, 186–8, 191–3, 200–4, 210, 214–15, 218–19
  betting strategy/ies 39, 82, 202–3
  robust strategy/ies 39, 82, 144, 200, 202–3
  semi-robust strategy/ies 39, 82, 202–3
strategy development 30, 35, 37–8, 155, 214–15
strategy mix 202
street festival/s 162
stresses 23–5, 27, 130
sustainable development 83, 138, 154, 188–90, 202–3
sustainable tourism xvii, 41, 211
sustainability xv, xviii, 28, 83, 95, 110, 118, 138, 156, 179, 203, 209, 216
Systems 9, 15, 23, 26–8, 33, 58, 64, 83, 129, 216–17

technology xvii, 5, 12, 24, 31, 41, 43, 53–7, 100, 106–7, 109–10, 114, 18, 22, 136–7, 209, 218
theoretical framework 15–16, 208
thinking 3, 7, 17–18, 28, 32–3, 36–8, 50, 58, 59, 69, 96, 129–30, 138, 144–5, 154, 156, 158–9, 214
  discursive thinking 32, 34
  dominant thinking 34, 38, 60, 69, 74, 76, 131
  holistic thinking 33
  linear mode of thinking 32
  nonlinear thinking 32
  outside-the-box thinking 38, 59, 138
  scenario thinking 10–11, 19, 35, 210, 214
threats 39, 42, 63–4, 81–2, 93, 104, 130, 133, 144, 155, 157, 160, 197, 202
tipping point(s) xiii, 25–6, 28, 46, 130, 137

tourism xiii, xv–xvi, 1–7, 22, 24, 33, 40, 42–3, 45, 47–50, 53–8, 60, 73, 79, 84–7, 89, 91–9, 104–5, 107, 110–11, 115, 117–19, 121, 123, 127–8, 147–54, 161, 179, 183, 187–9, 191–8, 200–2, 204, 206, 208–19
  inbound tourism 6, 62, 147–57, 161, 211
  international xv–xvi, 4, 7, 9, 47, 49, 86, 91–2, 94–5, 106–8, 111–12, 115–16, 119–20, 127, 139, 147–8, 150–2, 154–5, 157–62, 164, 169, 172, 177–9, 182, 184–5, 187, 196–8, 202, 208–10, 213
  overnight stay tourism 149–52, 155, 161
tourism futures xiii–xv, 1, 3–6, 12, 26, 40–1, 59, 86–7, 98–101, 129, 148, 163, 186, 207–8, 210, 212, 216–18
tourism life cycle 207
tourism theory 15
transformation 16–18, 24, 53–4, 136–7, 168–9
transitions 114, 122, 215–16
transport 86, 90, 108, 112, 116, 118, 120, 150–2, 154, 158–60, 187, 191–4, 196–8, 201–2, 217
travel flows 187
trends 4, 28, 32, 40–1, 43, 47–9, 54–5, 58, 61, 65–6, 72, 85, 100–3, 129, 133, 145, 147, 149, 153, 155
trends analysis xiii, 10, 32, 41, 47
trilemma, trilemma triangle 72–3, 212
truth claims 16–17
typology 12, 215

UN World Tourism Organisation (UNWTO) 210
uncertainty 10–12, 18, 32–3, 36, 40–1, 55, 58–9, 64, 69–72, 75, 85–7, 89–90, 93, 96, 102–3, 136–8, 155, 168, 172, 174, 189, 218
unique selling points (USPs) 171
urban tourism 7, 188–90, 202

van der Heijden, Kees 19, 35, 132
visioning 30
VisitScotland xiii, xv, 4, 40, 45–6, 57
visitor pressure 7, 149, 152, 186–9, 194, 200, 202–3, 211
volatility 57
Voros, Joseph 59, 60
voting, dot voting 71, 83, 102, 104, 168
VUCA 2, 39, 58, 217–18
vulnerability 95, 121, 123, 159

water management 114, 122
weak signals 24, 133

what if 216–17
Whitney, Diana 167
wildcards 84, 130
Williams, Nigel L. 162
wind-tunnel test 188
Wolfram, Gernot 187
workshop 10, 18, 50–2, 55, 65–7, 75–6, 80, 83, 102, 104, 123, 149, 165–6, 169–72, 183–4, 188–9, 195, 200

World Café 104, 124
World Futures Studies Federation (WFSF) 33–4
worldview 33, 156
worst-case scenario 29

Yeoman, Ian xiii, xv–xvi, 3–4, 8–13, 15–16, 18, 20, 40–1, 44–9, 51, 53–5, 57, 91, 217–18

For Product Safety Concerns and Information please contact our EU Authorised Representative:

Easy Access System Europe

Mustamäe tee 50

10621 Tallinn

Estonia

gpsr.requests@easproject.com

www.ingramcontent.com/pod-product-compliance
Ingram Content Group UK Ltd.
Pitfield, Milton Keynes, MK11 3LW, UK
UKHW021835140426

5217IPUK00021B/1471